Service Life and Durability of Reinforced Concrete Structures

RILEM BOOKSERIES

Volume 17

RILEM, The International Union of Laboratories and Experts in Construction Materials, Systems and Structures, founded in 1947, is a non-governmental scientific association whose goal is to contribute to progress in the construction sciences, techniques and industries, essentially by means of the communication it fosters between research and practice. RILEM's focus is on construction materials and their use in building and civil engineering structures, covering all phases of the building process from manufacture to use and recycling of materials. More information on RILEM and its previous publications can be found on www.RILEM.net.

More information about this series at http://www.springer.com/series/8781

Carmen Andrade · Joost Gulikers
Elisabeth Marie-Victoire
Editors

Service Life and Durability of Reinforced Concrete Structures

Selected Papers of the 8th International RILEM PhD Workshop held in Marne-la-Vallée, France, September 26–27, 2016

Springer

Editors
Carmen Andrade
CIMNE: International Center for Numerical Methods in Engineering
Universitat Politècnica de Catalunya (UPC)
Barcelona, Spain

Elisabeth Marie-Victoire
Laboratoire de Recherche des Monuments Historiques
Champs-sur-Marne, France

Joost Gulikers
Ministry of Infrastructure and Water Management, Rijkswaterstaat-GPO
Utrecht, The Netherlands

ISSN 2211-0844 ISSN 2211-0852 (electronic)
RILEM Bookseries
ISBN 978-3-030-07969-7 ISBN 978-3-319-90236-4 (eBook)
https://doi.org/10.1007/978-3-319-90236-4

Softcover re-print of the Hardcover 1st edition 2019

This Springer imprint is published by the registered company Springer Nature Switzerland AG
The registered company address is: Gewerbestrasse 11, 6330 Cham, Switzerland

Preface

The book contains selected papers of the works presented at the 8th International PhD Student Workshop on *Service Life and Durability of Reinforced Concrete Structures*, held at Marne-la-Vallée, France, on September 26 and 27, 2016, which was hosted by the LRMH (Laboratoire de Recherche des Monuments Historiques).

The RILEM PhD workshops on service life have a history which dates back to 2003 when Jan-Magnus Østvik, student of the Department of Civil Engineering supervised by Ø. Vennesland from the Technical University of Trondheim, Norway, was staying at the Institute of Construction Sciences of Spain with Dr. Carmen Andrade. J. Østvik took the initiative to organize a series of workshops related to reinforcement corrosion and service life, the first being at Madrid in 2003 and at Lofoten, Norway, in 2005. A major boost for the workshops was introduced through the activities of RILEM TC-213 MAI "*Model assisted integral service life prediction of steel reinforced concrete structures with respect to corrosion induced damage,*" with the main focus of this TC on analyzing service life models related to reinforcement corrosion. To date, successful workshops have been organized in Madrid-2007, Zagreb-2008, Guimarães-2009, Madrid-2010, Espoo-2012, Delft-2013, and Madrid-2014. All meetings have been under the auspices of RILEM EAC (Educational Activities Committee), with the aim to bring together young researchers in the field of durability of concrete. The workshops are always intended exclusively for PhD students who work at the frontier of knowledge in the field of durability of reinforced concrete.

Reinforced concrete structures may prove to be very durable; however, their gradual degradation over time impairing both serviceability and structural safety is still a matter of great practical concern in view of the economic consequences for assessment, maintenance, and repair. Corrosion of steel reinforcement is considered to be the most detrimental process responsible for structural deterioration. Many studies are in progress to develop a comprehensive engineering approach for assessment of the initiation and the propagation period in both uncracked and cracked concrete; however, essential aspects still remain controversial. In addition, the interaction between different degradation mechanisms such as ASR, acid or sulfate, and frost attack requires further understanding.

The present book collects the advanced lines of research developed in several universities to contribute to the basic understanding of the complex corrosion and degradation processes. It is grouped into the following subjects: (a) concrete properties and initiation period processes, (b) techniques of characterization of corrosion and degradation, (c) behavior of corroded structures, and (d) corrosion prevention and repair techniques. They represent examples of areas of research still needing development.

Barcelona, Spain Carmen Andrade
Utrecht, The Netherlands Joost Gulikers
Champs-sur-Marne, France Elisabeth Marie-Victoire

Contents

RILEM Publications

The following list is presenting the global offer of RILEM Publications, sorted by series. Each publication is available in printed version and/or in online version.

RILEM PROCEEDINGS (PRO)

PRO 1: Durability of High Performance Concrete (ISBN: 2-912143-03-9; e-ISBN: 2-351580-12-5; e-ISBN: 2351580125); *Ed. H. Sommer*

PRO 2: Chloride Penetration into Concrete (ISBN: 2-912143-00-04; e-ISBN: 2912143454); *Eds. L.-O. Nilsson and J.-P. Ollivier*

PRO 3: Evaluation and Strengthening of Existing Masonry Structures (ISBN: 2-912143-02-0; e-ISBN: 2351580141); *Eds. L. Binda and C. Modena*

PRO 4: Concrete: From Material to Structure (ISBN: 2-912143-04-7; e-ISBN: 2351580206); *Eds. J.-P. Bournazel and Y. Malier*

PRO 5: The Role of Admixtures in High Performance Concrete (ISBN: 2-912143-05-5; e-ISBN: 2351580214); *Eds. J. G. Cabrera and R. Rivera-Villarreal*

PRO 6: High Performance Fiber Reinforced Cement Composites—HPFRCC 3 (ISBN: 2-912143-06-3; e-ISBN: 2351580222); *Eds. H. W. Reinhardt and A. E. Naaman*

PRO 7: 1st International RILEM Symposium on Self-Compacting Concrete (ISBN: 2-912143-09-8; e-ISBN: 2912143721); *Eds. Å. Skarendahl and Ö. Petersson*

PRO 8: International RILEM Symposium on Timber Engineering (ISBN: 2-912143-10-1; e-ISBN: 2351580230); *Ed. L. Boström*

PRO 9: 2nd International RILEM Symposium on Adhesion between Polymers and Concrete ISAP'99 (ISBN: 2-912143-11-X; e-ISBN: 2351580249); *Eds. Y. Ohama and M. Puterman*

PRO 10: 3rd International RILEM Symposium on Durability of Building and Construction Sealants (ISBN: 2-912143-13-6; e-ISBN: 2351580257); *Eds. A. T. Wolf*

PRO 11: 4th International RILEM Conference on Reflective Cracking in Pavements (ISBN: 2-912143-14-4; e-ISBN: 2351580265); *Eds. A. O. Abd El Halim, D. A. Taylor and El H. H. Mohamed*

PRO 12: International RILEM Workshop on Historic Mortars: Characteristics and Tests (ISBN: 2-912143-15-2; e-ISBN: 2351580273); *Eds. P. Bartos, C. Groot and J. J. Hughes*

PRO 13: 2nd International RILEM Symposium on Hydration and Setting (ISBN: 2-912143-16-0; e-ISBN: 2351580281); *Ed. A. Nonat*

PRO 14: Integrated Life-Cycle Design of Materials and Structures—ILCDES 2000 (ISBN: 951-758-408-3; e-ISBN: 235158029X); (ISSN: 0356-9403); *Ed. S. Sarja*

PRO 15: Fifth RILEM Symposium on Fibre-Reinforced Concretes (FRC)—BEFIB'2000 (ISBN: 2-912143-18-7; e-ISBN: 291214373X); *Eds. P. Rossi and G. Chanvillard*

PRO 16: Life Prediction and Management of Concrete Structures (ISBN: 2-912143-19-5; e-ISBN: 2351580303); *Ed. D. Naus*

PRO 17: Shrinkage of Concrete—Shrinkage 2000 (ISBN: 2-912143-20-9; e-ISBN: 2351580311); *Eds. V. Baroghel-Bouny and P.-C. Aïtcin*

PRO 18: Measurement and Interpretation of the On-Site Corrosion Rate (ISBN: 2-912143-21-7; e-ISBN: 235158032X); *Eds. C. Andrade, C. Alonso, J. Fullea, J. Polimon and J. Rodriguez*

PRO 19: Testing and Modelling the Chloride Ingress into Concrete (ISBN: 2-912143-22-5; e-ISBN: 2351580338); *Eds. C. Andrade and J. Kropp*

PRO 20: 1st International RILEM Workshop on Microbial Impacts on Building Materials (CD 02) (e-ISBN 978-2-35158-013-4); *Ed. M. Ribas Silva*

PRO 21: International RILEM Symposium on Connections between Steel and Concrete (ISBN: 2-912143-25-X; e-ISBN: 2351580346); *Ed. R. Eligehausen*

PRO 22: International RILEM Symposium on Joints in Timber Structures (ISBN: 2-912143-28-4; e-ISBN: 2351580354); *Eds. S. Aicher and H.-W. Reinhardt*

PRO 23: International RILEM Conference on Early Age Cracking in Cementitious Systems (ISBN: 2-912143-29-2; e-ISBN: 2351580362); *Eds. K. Kovler and A. Bentur*

PRO 24: 2nd International RILEM Workshop on Frost Resistance of Concrete (ISBN: 2-912143-30-6; e-ISBN: 2351580370); *Eds. M. J. Setzer, R. Auberg and H.-J. Keck*

PRO 25: International RILEM Workshop on Frost Damage in Concrete (ISBN: 2-912143-31-4; e-ISBN: 2351580389); *Eds. D. J. Janssen, M. J. Setzer and M. B. Snyder*

PRO 26: International RILEM Workshop on On-Site Control and Evaluation of Masonry Structures (ISBN: 2-912143-34-9; e-ISBN: 2351580141); *Eds. L. Binda and R. C. de Vekey*

PRO 27: International RILEM Symposium on Building Joint Sealants (CD03; e-ISBN: 235158015X); *Ed. A. T. Wolf*

PRO 28: 6th International RILEM Symposium on Performance Testing and Evaluation of Bituminous Materials—PTEBM'03 (ISBN: 2-912143-35-7; e-ISBN: 978-2-912143-77-8); *Ed. M. N. Partl*

PRO 29: 2nd International RILEM Workshop on Life Prediction and Ageing Management of Concrete Structures (ISBN: 2-912143-36-5; e-ISBN: 2912143780); *Ed. D. J. Naus*

PRO 30: 4th International RILEM Workshop on High Performance Fiber Reinforced Cement Composites—HPFRCC 4 (ISBN: 2-912143-37-3; e-ISBN: 2912143799); *Eds. A. E. Naaman and H. W. Reinhardt*

PRO 31: International RILEM Workshop on Test and Design Methods for Steel Fibre Reinforced Concrete: Background and Experiences (ISBN: 2-912143-38-1; e-ISBN: 2351580168); *Eds. B. Schnütgen and L. Vandewalle*

PRO 32: International Conference on Advances in Concrete and Structures 2 vol. (ISBN (set): 2-912143-41-1; e-ISBN: 2351580176); *Eds. Ying-shu Yuan, Surendra P. Shah and Heng-lin Lü*

PRO 33: 3rd International Symposium on Self-Compacting Concrete (ISBN: 2-912143-42-X; e-ISBN: 2912143713); *Eds. Ó. Wallevik and I. Níelsson*

PRO 34: International RILEM Conference on Microbial Impact on Building Materials (ISBN: 2-912143-43-8; e-ISBN: 2351580184); *Ed. M. Ribas Silva*

PRO 35: International RILEM TC 186-ISA on Internal Sulfate Attack and Delayed Ettringite Formation (ISBN: 2-912143-44-6; e-ISBN: 2912143802); *Eds. K. Scrivener and J. Skalny*

PRO 36: International RILEM Symposium on Concrete Science and Engineering—A Tribute to Arnon Bentur (ISBN: 2-912143-46-2; e-ISBN: 2912143586); *Eds. K. Kovler, J. Marchand, S. Mindess and J. Weiss*

PRO 37: 5th International RILEM Conference on Cracking in Pavements—Mitigation, Risk Assessment and Prevention (ISBN: 2-912143-47-0; e-ISBN: 2912143764); *Eds. C. Petit, I. Al-Qadi and A. Millien*

PRO 38: 3rd International RILEM Workshop on Testing and Modelling the Chloride Ingress into Concrete (ISBN: 2-912143-48-9; e-ISBN: 2912143578); *Eds. C. Andrade and J. Kropp*

PRO 39: 6th International RILEM Symposium on Fibre-Reinforced Concretes—BEFIB 2004 (ISBN: 2-912143-51-9; e-ISBN: 2912143748); *Eds. M. Di Prisco, R. Felicetti and G. A. Plizzari*

PRO 40: International RILEM Conference on the Use of Recycled Materials in Buildings and Structures (ISBN: 2-912143-52-7; e-ISBN: 2912143756); *Eds. E. Vázquez, Ch. F. Hendriks and G. M. T. Janssen*

PRO 41: RILEM International Symposium on Environment-Conscious Materials and Systems for Sustainable Development (ISBN: 2-912143-55-1; e-ISBN: 2912143640); *Eds. N. Kashino and Y. Ohama*

PRO 42: SCC'2005—China: 1st International Symposium on Design, Performance and Use of Self-Consolidating Concrete (ISBN: 2-912143-61-6; e-ISBN: 2912143624); *Eds. Zhiwu Yu, Caijun Shi, Kamal Henri Khayat and Youjun Xie*

PRO 43: International RILEM Workshop on Bonded Concrete Overlays (e-ISBN: 2-912143-83-7); *Eds. J. L. Granju and J. Silfwerbrand*

PRO 44: 2nd International RILEM Workshop on Microbial Impacts on Building Materials (CD11) (e-ISBN: 2-912143-84-5); *Ed. M. Ribas Silva*

PRO 45: 2nd International Symposium on Nanotechnology in Construction, Bilbao (ISBN: 2-912143-87-X; e-ISBN: 2912143888); *Eds. Peter J. M. Bartos, Yolanda de Miguel and Antonio Porro*

PRO 46: ConcreteLife'06—International RILEM-JCI Seminar on Concrete Durability and Service Life Planning: Curing, Crack Control, Performance in Harsh Environments (ISBN: 2-912143-89-6; e-ISBN: 291214390X); *Ed. K. Kovler*

PRO 47: International RILEM Workshop on Performance Based Evaluation and Indicators for Concrete Durability (ISBN: 978-2-912143-95-2; e-ISBN: 9782912143969); *Eds. V. Baroghel-Bouny, C. Andrade, R. Torrent and K. Scrivener*

PRO 48: 1st International RILEM Symposium on Advances in Concrete through Science and Engineering (e-ISBN: 2-912143-92-6); *Eds. J. Weiss, K. Kovler, J. Marchand, and S. Mindess*

PRO 49: International RILEM Workshop on High Performance Fiber Reinforced Cementitious Composites in Structural Applications (ISBN: 2-912143-93-4; e-ISBN: 2912143942); *Eds. G. Fischer and V. C. Li*

PRO 50: 1st International RILEM Symposium on Textile Reinforced Concrete (ISBN: 2-912143-97-7; e-ISBN: 2351580087); *Eds. Josef Hegger, Wolfgang Brameshuber and Norbert Will*

PRO 51: 2nd International Symposium on Advances in Concrete through Science and Engineering (ISBN: 2-35158-003-6; e-ISBN: 2-35158-002-8); *Eds. J. Marchand, B. Bissonnette, R. Gagné, M. Jolin and F. Paradis*

PRO 52: Volume Changes of Hardening Concrete: Testing and Mitigation (ISBN: 2-35158-004-4; e-ISBN: 2-35158-005-2); *Eds. O. M. Jensen, P. Lura and K. Kovler*

PRO 53: High Performance Fiber Reinforced Cement Composites—HPFRCC5 (ISBN: 978-2-35158-046-2; e-ISBN: 978-2-35158-089-9); *Eds. H. W. Reinhardt and A. E. Naaman*

PRO 54: 5th International RILEM Symposium on Self-Compacting Concrete (ISBN: 978-2-35158-047-9; e-ISBN: 978-2-35158-088-2); *Eds. G. De Schutter and V. Boel*

PRO 55: International RILEM Symposium Photocatalysis, Environment and Construction Materials (ISBN: 978-2-35158-056-1; e-ISBN: 978-2-35158-057-8); *Eds. P. Baglioni and L. Cassar*

PRO 56: International RILEM Workshop on Integral Service Life Modelling of Concrete Structures (ISBN 978-2-35158-058-5; e-ISBN: 978-2-35158-090-5); *Eds. R. M. Ferreira, J. Gulikers and C. Andrade*

PRO 57: RILEM Workshop on Performance of cement-based materials in aggressive aqueous environments (e-ISBN: 978-2-35158-059-2); *Ed. N. De Belie*

PRO 58: International RILEM Symposium on Concrete Modelling—CONMOD'08 (ISBN: 978-2-35158-060-8; e-ISBN: 978-2-35158-076-9); *Eds. E. Schlangen and G. De Schutter*

PRO 59: International RILEM Conference on On Site Assessment of Concrete, Masonry and Timber Structures—SACoMaTiS 2008 (ISBN set: 978-2-35158-061-5; e-ISBN: 978-2-35158-075-2); *Eds. L. Binda, M. di Prisco and R. Felicetti*

PRO 60: Seventh RILEM International Symposium on Fibre Reinforced Concrete: Design and Applications—BEFIB 2008 (ISBN: 978-2-35158-064-6; e-ISBN: 978-2-35158-086-8); *Ed. R. Gettu*

PRO 61: 1st International Conference on Microstructure Related Durability of Cementitious Composites 2 vol., (ISBN: 978-2-35158-065-3; e-ISBN: 978-2-35158-084-4); *Eds. W. Sun, K. van Breugel, C. Miao, G. Ye and H. Chen*

PRO 62: NSF/ RILEM Workshop: In-situ Evaluation of Historic Wood and Masonry Structures (e-ISBN: 978-2-35158-068-4); *Eds. B. Kasal, R. Anthony and M. Drdácký*

PRO 63: Concrete in Aggressive Aqueous Environments: Performance, Testing and Modelling, 2 vol., (ISBN: 978-2-35158-071-4; e-ISBN: 978-2-35158-082-0); *Eds. M. G. Alexander and A. Bertron*

PRO 64: Long Term Performance of Cementitious Barriers and Reinforced Concrete in Nuclear Power Plants and Waste Management—NUCPERF 2009 (ISBN: 978-2-35158-072-1; e-ISBN: 978-2-35158-087-5); *Eds. V. L'Hostis, R. Gens, C. Gallé*

PRO 65: Design Performance and Use of Self-consolidating Concrete—SCC'2009 (ISBN: 978-2-35158-073-8; e-ISBN: 978-2-35158-093-6); *Eds. C. Shi, Z. Yu, K. H. Khayat and P. Yan*

PRO 66: 2nd International RILEM Workshop on Concrete Durability and Service Life Planning—ConcreteLife'09 (ISBN: 978-2-35158-074-5; ISBN: 978-2-35158-074-5); *Ed. K. Kovler*

PRO 67: Repairs Mortars for Historic Masonry (e-ISBN: 978-2-35158-083-7); *Ed. C. Groot*

PRO 68: Proceedings of the 3rd International RILEM Symposium on 'Rheology of Cement Suspensions such as Fresh Concrete (ISBN 978-2-35158-091-2; e-ISBN: 978-2-35158-092-9); *Eds. O. H. Wallevik, S. Kubens and S. Oesterheld*

PRO 69: 3rd International PhD Student Workshop on 'Modelling the Durability of Reinforced Concrete (ISBN: 978-2-35158-095-0); *Eds. R. M. Ferreira, J. Gulikers and C. Andrade*

PRO 70: 2nd International Conference on 'Service Life Design for Infrastructure' (ISBN set: 978-2-35158-096-7, e-ISBN: 978-2-35158-097-4); *Ed. K. van Breugel, G. Ye and Y. Yuan*

PRO 71: Advances in Civil Engineering Materials—The 50-year Teaching Anniversary of Prof. Sun Wei' (ISBN: 978-2-35158-098-1; e-ISBN: 978-2-35158-099-8); *Eds. C. Miao, G. Ye, and H. Chen*

PRO 72: First International Conference on 'Advances in Chemically-Activated Materials—CAM'2010' (2010), 264 pp, ISBN: 978-2-35158-101-8; e-ISBN: 978-2-35158-115-5, *Eds. Caijun Shi and Xiaodong Shen*

PRO 73: 2nd International Conference on 'Waste Engineering and Management—ICWEM 2010' (2010), 894 pp, ISBN: 978-2-35158-102-5; e-ISBN: 978-2-35158-103-2, *Eds. J. Zh. Xiao, Y. Zhang, M. S. Cheung and R. Chu*

PRO 74: International RILEM Conference on 'Use of Superabsorsorbent Polymers and Other New Addditives in Concrete' (2010) 374 pp., ISBN: 978-2-35158-104-9; e-ISBN: 978-2-35158-105-6; *Eds. O. M. Jensen, M. T. Hasholt, and S. Laustsen*

PRO 75: International Conference on 'Material Science—2nd ICTRC—Textile Reinforced Concrete—Theme 1' (2010) 436 pp., ISBN: 978-2-35158-106-3; e-ISBN: 978-2-35158-107-0; *Ed. W. Brameshuber*

PRO 76: International Conference on 'Material Science—HetMat—Modelling of Heterogeneous Materials—Theme 2' (2010) 255 pp., ISBN: 978-2-35158-108-7; e-ISBN: 978-2-35158-109-4; *Ed. W. Brameshuber*

PRO 77: International Conference on 'Material Science—AdIPoC—Additions Improving Properties of Concrete—Theme 3' (2010) 459 pp., ISBN: 978-2-35158-110-0; e-ISBN: 978-2-35158-111-7; *Ed. W. Brameshuber*

PRO 78: 2nd Historic Mortars Conference and RILEM TC 203-RHM Final Workshop—HMC2010 (2010) 1416 pp., e-ISBN: 978-2-35158-112-4; *Eds. J. Válek, C. Groot, and J. J. Hughes*

PRO 79: International RILEM Conference on Advances in Construction Materials Through Science and Engineering (2011) 213 pp., ISBN: 978-2-35158-116-2, e-ISBN: 978-2-35158-117-9; *Eds. Christopher Leung and K. T. Wan*

PRO 80: 2nd International RILEM Conference on Concrete Spalling due to Fire Exposure (2011) 453 pp., ISBN: 978-2-35158-118-6, e-ISBN: 978-2-35158-119-3; *Eds. E. A. B. Koenders and F. Dehn*

PRO 81: 2nd International RILEM Conference on Strain Hardening Cementitious Composites (SHCC2-Rio) (2011) 451 pp., ISBN: 978-2-35158-120-9, e-ISBN: 978-2-35158-121-6; *Eds. R.D. Toledo Filho, F. A. Silva, E. A. B. Koenders and E. M. R. Fairbairn*

PRO 82: 2nd International RILEM Conference on Progress of Recycling in the Built Environment (2011) 507 pp., e-ISBN: 978-2-35158-122-3; *Eds. V. M. John, E. Vazquez, S. C. Angulo and C. Ulsen*

PRO 83: 2nd International Conference on Microstructural-related Durability of Cementitious Composites (2012) 250 pp., ISBN: 978-2-35158-129-2; e-ISBN: 978-2-35158-123-0; *Eds. G. Ye, K. van Breugel, W. Sun and C. Miao*

PRO 84: CONSEC13—Seventh International Conference on Concrete under Severe Conditions—Environment and Loading (2013) 1930 pp., ISBN: 978-2-35158-124-7; e-ISBN: 978-2- 35158-134-6; *Eds. Z.J. Li, W. Sun, C. W. Miao, K. Sakai, O. E. Gjorv & N. Banthia*

PRO 85: RILEM-JCI International Workshop on Crack Control of Mass Concrete and Related issues concerning Early-Age of Concrete Structures—ConCrack 3—Control of Cracking in Concrete Structures 3 (2012) 237 pp., ISBN: 978-2-35158-125-4; e-ISBN: 978-2-35158-126-1; *Eds. F. Toutlemonde and J.-M. Torrenti*

PRO 86: International Symposium on Life Cycle Assessment and Construction (2012) 414 pp., ISBN: 978-2-35158-127-8, e-ISBN: 978-2-35158-128-5; *Eds. A. Ventura and C. de la Roche*

PRO 87: UHPFRC 2013—RILEM-fib-AFGC International Symposium on Ultra-High Performance Fibre-Reinforced Concrete (2013), ISBN: 978-2-35158-130-8, e-ISBN: 978-2-35158-131-5; *Eds. F. Toutlemonde*

PRO 88: 8th RILEM International Symposium on Fibre Reinforced Concrete (2012) 344 pp., ISBN: 978-2-35158-132-2, e-ISBN: 978-2-35158-133-9; *Eds. Joaquim A. O. Barros*

PRO 89: RILEM International workshop on performance-based specification and control of concrete durability (2014) 678 pp, ISBN: 978-2-35158-135-3, e-ISBN: 978-2-35158-136-0; *Eds. D. Bjegović, H. Beushausen and M. Serdar*

PRO 90: 7th RILEM International Conference on Self-Compacting Concrete and of the 1st RILEM International Conference on Rheology and Processing of Construction Materials (2013) 396 pp, ISBN: 978-2-35158-137-7, e-ISBN: 978-2-35158-138-4; *Eds. Nicolas Roussel and Hela Bessaies-Bey*

PRO 91: CONMOD 2014—RILEM International Symposium on Concrete Modelling (2014), ISBN: 978-2-35158-139-1; e-ISBN: 978-2-35158-140-7; *Eds. Kefei Li, Peiyu Yan and Rongwei Yang*

PRO 92: CAM 2014—2nd International Conference on advances in chemically-activated materials (2014) 392 pp., ISBN: 978-2-35158-141-4; e-ISBN: 978-2-35158-142-1; *Eds. Caijun Shi and Xiadong Shen*

PRO 93: SCC 2014—3rd International Symposium on Design, Performance and Use of Self-Consolidating Concrete (2014) 438 pp., ISBN: 978-2-35158-143-8; e-ISBN: 978-2-35158-144-5; *Eds. Caijun Shi, Zhihua Ou, Kamal H. Khayat*

PRO 94 (online version): HPFRCC-7—7th RILEM conference on High performance fiber reinforced cement composites (2015), e-ISBN: 978-2-35158-146-9; *Eds. H. W. Reinhardt, G. J. Parra-Montesinos, H. Garrecht*

PRO 95: International RILEM Conference on Application of superabsorbent polymers and other new admixtures in concrete construction (2014), ISBN: 978-2-35158-147-6; e-ISBN: 978-2-35158-148-3; *Eds. Viktor Mechtcherine, Christof Schroefl*

PRO 96 (online version): XIII DBMC: XIII International Conference on Durability of Building Materials and Components(2015), e-ISBN: 978-2-35158-149-0; *Eds. M. Quattrone, V. M. John*

PRO 97: SHCC3—3rd International RILEM Conference on Strain Hardening Cementitious Composites (2014), ISBN: 978-2-35158-150-6; e-ISBN: 978-2-35158-151-3; *Eds. E. Schlangen, M. G. Sierra Beltran, M. Lukovic, G. Ye*

PRO 98: FERRO-11—11th International Symposium on Ferrocement and 3rd ICTRC—International Conference on Textile Reinforced Concrete (2015), ISBN: 978-2-35158-152-0; e-ISBN: 978-2-35158-153-7; *Ed. W. Brameshuber*

PRO 99 (online version): ICBBM 2015—1st International Conference on Bio-Based Building Materials (2015), e-ISBN: 978-2-35158-154-4; *Eds. S. Amziane, M. Sonebi*

PRO 100: SCC16—RILEM Self-Consolidating Concrete Conference (2016), ISBN: 978-2-35158-156-8; e-ISBN: 978-2-35158-157-5; *Ed. Kamal H. Kayat*

PRO 101 (online version): III Progress of Recycling in the Built Environment (2015), e-ISBN: 978-2-35158-158-2; *Eds I. Martins, C. Ulsen and S. C. Angulo*

PRO 102 (online version): RILEM Conference on Microorganisms-Cementitious Materials Interactions (2016), e-ISBN: 978-2-35158-160-5; *Eds. Alexandra Bertron, Henk Jonkers, Virginie Wiktor*

PRO 103 (online version): ACESC'16—Advances in Civil Engineering and Sustainable Construction (2016), e-ISBN: 978-2-35158-161-2; *Eds. T.Ch. Madhavi, G. Prabhakar, Santhosh Ram and P. M. Rameshwaran*

PRO 104 (online version): SSCS'2015—Numerical Modeling—Strategies for Sustainable Concrete Structures (2015), e-ISBN: 978-2-35158-162-9

PRO 105: 1st International Conference on UHPC Materials and Structures (2016), ISBN: 978-2-35158-164-3, e-ISBN: 978-2-35158-165-0

PRO 106: AFGC-ACI-fib-RILEM International Conference on Ultra-High-Performance Fibre-Reinforced Concrete—UHPFRC 2017 (2017), ISBN: 978-2-35158-166-7, e-ISBN: 978-2-35158-167-4; *Eds. François Toutlemonde & Jacques Resplendino*

PRO 107 (online version): XIV DBMC—14th International Conference on Durability of Building Materials and Components (2017), e-ISBN: 978-2-35158-159-9; *Eds. Geert De Schutter, Nele De Belie, Arnold Janssens, Nathan Van Den Bossche*

PRO 108: MSSCE 2016—Innovation of Teaching in Materials and Structures (2016), ISBN: 978-2-35158-178-0, e-ISBN: 978-2-35158-179-7; *Ed. Per Goltermann*

PRO 109 (2 volumes): MSSCE 2016—Service Life of Cement-Based Materials and Structures (2016), ISBN Vol. 1: 978-2-35158-170-4, Vol. 2: 978-2-35158-171-4, Set Vol. 1&2: 978-2-35158-172-8, e-ISBN : 978-2-35158-173-5; *Eds. Miguel Azenha, Ivan Gabrijel, Dirk Schlicke, Terje Kanstad and Ole Mejlhede Jensen*

PRO 110: MSSCE 2016—Historical Masonry (2016), ISBN: 978-2-35158-178-0, e-ISBN: 978-2-35158-179-7; *Eds. Inge Rörig-Dalgaard and Ioannis Ioannou*

PRO 111: MSSCE 2016—Electrochemistry in Civil Engineering (2016), ISBN: 978-2-35158-176-6, e-ISBN: 978-2-35158-177-3; *Ed. Lisbeth M. Ottosen*

PRO 112: MSSCE 2016—Moisture in Materials and Structures (2016), ISBN: 978-2-35158-178-0, e-ISBN: 978-2-35158-179-7; *Eds. Kurt Kielsgaard Hansen, Carsten Rode and Lars-Olof Nilsson*

PRO 113: MSSCE 2016—Concrete with Supplementary Cementitious Materials (2016), ISBN: 978-2-35158-178-0, e-ISBN: 978-2-35158-179-7; *Eds. Ole Mejlhede Jensen, Konstantin Kovler and Nele De Belie*

PRO 114: MSSCE 2016—Frost Action in Concrete (2016), ISBN: 978-2-35158-182-7, e-ISBN: 978-2-35158-183-4; *Eds. Marianne Tange Hasholt, Katja Fridh and R. Doug Hooton*

PRO 115: MSSCE 2016—Fresh Concrete (2016), ISBN: 978-2-35158-184-1, e-ISBN: 978-2-35158-185-8; *Eds. Lars N. Thrane, Claus Pade, Oldrich Svec and Nicolas Roussel*

PRO 116: BEFIB 2016—9th RILEM International Symposium on Fiber Reinforced Concrete (2016), ISBN: 978-2-35158-187-2, e-ISBN: 978-2-35158-186-5; *Eds. N. Banthia, M. di Prisco and S. Soleimani-Dashtaki*

PRO 117: 3rd International RILEM Conference on Microstructure Related Durability of Cementitious Composites (2016), ISBN: 978-2-35158-188-9, e-ISBN: 978-2-35158-189-6; *Eds. Changwen Miao, Wei Sun, Jiaping Liu, Huisu Chen, Guang Ye and Klaas van Breugel*

PRO 118 (4 volumes): International Conference on Advances in Construction Materials and Systems (2017), ISBN Set: 978-2-35158-190-2, Vol. 1: 978-2-35158-193-3, Vol. 2: 978-2-35158-194-0, Vol. 3: ISBN:978-2-35158-195-7, Vol. 4: ISBN:978-2-35158-196-4, e-ISBN: 978-2-35158-191-9; *Ed. Manu Santhanam*

PRO 119 (online version): ICBBM 2017—Second International RILEM Conference on Bio-based Building Materials, (2017), e-ISBN: 978-2-35158-192-6; *Ed. Sofiane Amziane*

PRO 120: 2nd International RILEM/COST Conference on Early Age Cracking and Serviceability in Cement-based Materials and Structures (EAC-02), 2017, ISBN: 978-2-35158-197-1, e-ISBN: 978-2-35158-198-8; *Eds. Dimitrios Aggelis and Stéphanie Staquet*

PRO 121: SynerCrete18, Interdisciplinary Approaches for Cement-based Materials and Structural Concrete: Synergizing Expertise and Bridging Scales of Space and Time, (2018), ISBN: 978-2-35158-202-2, e-ISBN: 978-2-35158-203-9; *Eds. Miguel Azenha, Dirk Schlicke, Farid Benboudjema, Agnieszka Knoppik*

RILEM REPORTS (REP)

Report 19: Considerations for Use in Managing the Aging of Nuclear Power Plant Concrete Structures (ISBN: 2-912143-07-1); *Ed. D. J. Naus*

Report 20: Engineering and Transport Properties of the Interfacial Transition Zone in Cementitious Composites (ISBN: 2-912143-08-X); *Eds. M. G. Alexander, G. Arliguie, G. Ballivy, A. Bentur and J. Marchand*

Report 21: Durability of Building Sealants (ISBN: 2-912143-12-8); *Ed. A. T. Wolf*

Report 22: Sustainable Raw Materials—Construction and Demolition Waste (ISBN: 2-912143-17-9); *Eds. C. F. Hendriks and H. S. Pietersen*

Report 23: Self-Compacting Concrete state-of-the-art report (ISBN: 2-912143-23-3); *Eds. Å. Skarendahl and Ö. Petersson*

Report 24: Workability and Rheology of Fresh Concrete: Compendium of Tests (ISBN: 2-912143-32-2); *Eds. P. J. M. Bartos, M. Sonebi and A. K. Tamimi*

Report 25: Early Age Cracking in Cementitious Systems (ISBN: 2-912143-33-0); *Ed. A. Bentur*

Report 26: Towards Sustainable Roofing (Joint Committee CIB/RILEM) (CD 07) (e-ISBN 978-2-912143-65-5); *Eds. Thomas W. Hutchinson and Keith Roberts*

Report 27: Condition Assessment of Roofs (Joint Committee CIB/RILEM) (CD 08) (e-ISBN 978-2-912143-66-2); *Ed. CIB W 83/RILEM TC166-RMS*

Report 28: Final report of RILEM TC 167-COM 'Characterisation of Old Mortars with Respect to Their Repair (ISBN: 978-2-912143-56-3); *Eds. C. Groot, G. Ashall and J. Hughes*

Report 29: Pavement Performance Prediction and Evaluation (PPPE): Interlaboratory Tests (e-ISBN: 2-912143-68-3); *Eds. M. Partl and H. Piber*

Report 30: Final Report of RILEM TC 198-URM 'Use of Recycled Materials' (ISBN: 2-912143-82-9; e-ISBN: 2-912143-69-1); *Eds. Ch. F. Hendriks, G. M. T. Janssen and E. Vázquez*

Report 31: Final Report of RILEM TC 185-ATC 'Advanced testing of cement-based materials during setting and hardening' (ISBN: 2-912143-81-0; e-ISBN: 2-912143-70-5); *Eds. H. W. Reinhardt and C. U. Grosse*

Report 32: Probabilistic Assessment of Existing Structures. A JCSS publication (ISBN 2-912143-24-1); *Ed. D. Diamantidis*

Report 33: State-of-the-Art Report of RILEM Technical Committee TC 184-IFE 'Industrial Floors' (ISBN 2-35158-006-0); *Ed. P. Seidler*

Report 34: Report of RILEM Technical Committee TC 147-FMB 'Fracture mechanics applications to anchorage and bond' Tension of Reinforced Concrete Prisms—Round Robin Analysis and Tests on Bond (e-ISBN 2-912143-91-8); *Eds. L. Elfgren and K. Noghabai*

Report 35: Final Report of RILEM Technical Committee TC 188-CSC 'Casting of Self Compacting Concrete' (ISBN 2-35158-001-X; e-ISBN: 2-912143-98-5); *Eds. Å. Skarendahl and P. Billberg*

Report 36: State-of-the-Art Report of RILEM Technical Committee TC 201-TRC 'Textile Reinforced Concrete' (ISBN 2-912143-99-3); *Ed. W. Brameshuber*

Report 37: State-of-the-Art Report of RILEM Technical Committee TC 192-ECM 'Environment-conscious construction materials and systems' (ISBN: 978-2-35158-053-0); *Eds. N. Kashino, D. Van Gemert and K. Imamoto*

Report 38: State-of-the-Art Report of RILEM Technical Committee TC 205-DSC 'Durability of Self-Compacting Concrete' (ISBN: 978-2-35158-048-6); *Eds. G. De Schutter and K. Audenaert*

Report 39: Final Report of RILEM Technical Committee TC 187-SOC 'Experimental determination of the stress-crack opening curve for concrete in tension' (ISBN 978-2-35158-049-3); *Ed. J. Planas*

Report 40: State-of-the-Art Report of RILEM Technical Committee TC 189-NEC 'Non-Destructive Evaluation of the Penetrability and Thickness of the Concrete Cover' (ISBN 978-2-35158-054-7); *Eds. R. Torrent and L. Fernández Luco*

Report 41: State-of-the-Art Report of RILEM Technical Committee TC 196-ICC 'Internal Curing of Concrete' (ISBN 978-2-35158-009-7); *Eds. K. Kovler and O. M. Jensen*

Report 42: 'Acoustic Emission and Related Non-destructive Evaluation Techniques for Crack Detection and Damage Evaluation in Concrete'—Final Report of RILEM Technical Committee 212-ACD (e-ISBN: 978-2-35158-100-1); *Ed. M. Ohtsu*

Report 45: Repair Mortars for Historic Masonry—State-of-the-Art Report of RILEM Technical Committee TC 203-RHM (e-ISBN: 978-2-35158-163-6); *Eds. Paul Maurenbrecher and Caspar Groot*

Report 46: *S*urface delamination of concrete industrial floors and other durability related aspects guide—Report of RILEM Technical Committee TC 268-SIF (e-ISBN: 978-2-35158-201-5); *Ed. Valerie Pollet*

Part I
Concrete Properties and Initiation Period Processes

Effect of LDH Nano-Flakes on the Mechanical and Transport Properties of Lightweight Concrete

Z. Y. Qu, Q. L. Yu and H. J. H. Brouwers

Abstract Ca–Al–NO_3 layered double hydroxides (LDHs) with its layer structure presents potential to inhibit the ion transport process in the host matrix due to the increased tortuosity. In this study, the transport properties of lightweight concrete containing Ca–Al–NO_3 layered double hydroxides (LDHs) are investigated. A co-precipitation method is applied to synthesize the Ca–Al–NO_3 LDHs and the effect of the synthesis environment on the size and particle shape is investigated. The synthesized Ca–Al–NO_3 LDHs are analytically characterized by XRD, SEM and FTIR analyses. A lightweight concrete (LWC) is designed applying the modified Andreasen and Andersen model and a natural expansive pumice as lightweight aggregates. The synthesized LDHs are applied to the LWC with a fixed volume amount but different sizes. The effect of the LDHs on the mechanical properties and transport properties of the designed LWC are investigated and furthermore the effect of the size of the LDHs on these properties is discussed.

1 Introduction

Lightweight concrete (LWC) has been used as a building material over 3000 years. Owing to the advantages such as low density, good thermal insulation and fire resistance, LWC has been widely studied as both structural and nonstructural material such as for long span bridges, high rise buildings, buildings where foundation conditions are poor, or highly specialized applications such as floating and offshore structures [1–4].

Z. Y. Qu (✉) · Q. L. Yu · H. J. H. Brouwers
Department of the Built Environment, Eindhoven University of Technology, P. O. Box 513, 5600 MB Eindhoven, The Netherlands
e-mail: Z.Qu@tue.nl

Z. Y. Qu · H. J. H. Brouwers
State Key Laboratory of Silicate Materials for Architectures, Wuhan University of Technology, Wuhan 430070, People's Republic of China

C. Andrade et al. (eds.), *Service Life and Durability of Reinforced Concrete Structures*, RILEM Bookseries 17, https://doi.org/10.1007/978-3-319-90236-4_1

Compared with the normal aggregates, lightweight aggregates (LWA) are more porous which would potentially increase the pore connectivity in the LWC. Hence, the overall permeability of the composite containing LWA will increase and the deleterious fluids may penetrate into the LWC more easily. In order to improve the durability of LWC, one efficient strategy is to add nanoparticles that have been used in cement-based materials already to enhance the durability performance associated with many other merits [1, 3]. The addition of these nano-particles can fill the voids in the cement paste, resulting in a lower porosity and thus reduce the water permeability of concrete [5]. Nano-silica has been used to enhance the compressive strength and reduce the overall permeability of hardened concrete due to its pozzolanic properties, which result in finer hydrated phases (C–S–H gel) and densified microstructure (nano-filler and anti-$Ca(OH)_2$-leaching effect) [1, 6–8]. Sanchez et al. reported that the addition of nano-silica, nano-$CaCO_3$ and nano-TiO_2 can decrease the porosity and therefore reduce the permeability of concrete to chloride [9]. The effect of nanomaterials on the mechanical and transport properties of lightweight concrete has also been investigated [8]. Zhang et al. found that both the resistance to water and chloride ions penetration and mechanical properties of LWC were enhanced after the addition of 1% nano-silica [10].

Recently, there has been rising interest to apply 2D nano-materials to increase the durability of concrete. The layered structure of 2D nano-materials is effective in inhibiting the transport processes for fluids in the concrete owing to the increased tortuous paths. Du et al. reported the use of graphene nano-platelet (GNP) in cement mortar and concrete to study the barrier effect of GNP on the transport properties [11, 12]. The addition of the random distributed GNP can enhance the tortuosity and decrease the chloride transport by 50% in both concrete and mortar. Although various volume fractions of the GNP have been investigated, the influence of the sizes of the flakes on the transport property of concrete is still unclear. Especially filler sizes (i.e. micro filler effect) strongly influence the concrete property. However, due to the relatively easy bending feature of the GNP, the mechanical properties of the hybrid mortar and concrete did not increase. LDHs are a class of synthetic anionic clays with a typical flake shape. Owing to the availability of facile synthetic methods as well as the structural characteristics, it is possible to prepare LDHs and LDH-based materials with various physical and chemical properties [11]. A simple and cost-effective route to prepare the LDH is co-precipitation method. In most of the studies, the synthesis of LDH compounds is realised at a high pH value ($\geq$10) for the co-precipitation of trivalent and divalent cations [13].

The present research aims at investigating the influence of nano-flakes sizes to the mechanical and transport properties of the LWC. Through the control of the pH value of the precursor solution, Ca–Al–NO_3 LDHs with different sizes are synthesized by using a co-precipitation method. The alteration of morphology and sizes of Ca–Al LDH prepared under different pH values are studied and discussed. Then a lightweight concrete is designed applying the modified Andreasen and Andersen model and natural expansive pumice as lightweight aggregates. Subsequently, the influences of the as prepared LDHs with different sizes at an addition of 1% by volume, on the mechanical properties and transport properties of concrete are investigated. Rapid Chloride Migration (RCM) experiments were applied to investigate the barrier performance of the LDH nano-flakes contained LWC, and the LWC exhibited an enhanced chloride resistance of up to 54%.

2 Experiments

2.1 Preparation and Characterization of the Synthetic Ca–Al–NO_3 LDHs

Hydrocalumite (Ca–Al–NO_3 LDHs) is synthesized by using the co-precipitation method as reported elsewhere [14]. In the present study, the pH value of the solution is changed to prepare LDH nano-platelets with different sizes. Calcium nitrate tetrahydrate ($Ca(NO_3)_2{\cdot}4H_2O$) and aluminum nitrate nonahydrate ($Al(NO_3)_3{\cdot}9H_2O$) are dissolved in 200 ml deionized water with a stoichiometric ratio of 2:1 (4/3 and 2/3 M) to give a 2 M solution. This solution is added into 200 ml sodium nitrate ($NaNO_3$) with a concentration of 2 M. The mixed solution is stirred vigorously with a magnetic stirrer for 2 h at room temperature (20 ± 1 °C). The pH of the solution is adjusted to 11 and 13 (monitored by a pH-meter) by adding 1 M sodium hydroxide (NaOH, Sigma-Aldrich, AR) solution. The precipitate is then filtered in a vacuum enhanced process and the obtained filter cake is washed with deionized water until the filtrate is free of soluble nitrates. The solid is then dried at 100 °C in an oven for 12 h.

X-ray diffractometry (XRD) analysis was performed by using a Cu tube (40 kV, 30 mA) with a scanning range from 5° to 65° 2θ, applying a step 0.02 and 5 s/step measuring time. The qualitative analysis was carried out by using the Diffrac^{plus} Software (Bruker AXS) and the PDF database of ICDD.

The morphological features of the Ca–Al–NO_3 LDHs and the morphology of the Ca–Al–NO_3 LDHs added in the mortars were observed with a FE-SEM (JOEL JSM-5600). The obtained SEM image was also used to calculate the sizes and thickness of the LDHs. The FT-IR spectra of the reaction products were collected using a PerkinElmer Frontier^{TM} MIR/FIR Spectrometers using the attenuated total reflection (ATR) method (GladiATR). All spectra were scanned 48 times from 4000 to 400 cm^{-1} at a resolution of 4 cm^{-1}.

The specific density was obtained by using a gas pycnometer (AccuPyc II 1340). The AccuPyc works by measuring the amount of displaced gas (helium). The very small helium molecules rapidly fill the tiniest pores of the sample and only the true solid phase of the sample displaces the gas. The pressures are measured upon filling the gas into the sample chamber and then discharging it into a second empty chamber, to compute the solid phase volume of the sample.

The specific surface area was recorded by Tristar II instrument and is calculated from the amount of N_2 gas adsorbed based on the N_2 molecular cross sectional area of 0.162 nm^2.

2.2 Preparation of the LWC

The cement used in this study is Blastfurnace Cement CEM III/A 52.5 N (ENCI, The Netherlands). The lightweight aggregates used here are commercially available products manufactured from natural expanded silicate in Germany. Limestone powder and fly ash are used as filler to adjust the powder amount in the mixture. The water/cement ratio is fixed at 0.4. A polycarboxylic ether based superplasticizer (SP) is used to adjust the workability of concrete. The LDH prepared under different pH values was added at the content of 1% by volume of concrete. In order to investigate the barrier effect of discrete distribution nano-flakes to, a 50/50% mixture of LDH-11 and LDH-13 is also used. The synthetic LDHs were mixed in the SP-water solution (50% water), first for 10 min by hand and then added in the mixing process to prepare the LWC.

The mixes of the cement-based lightweight composite (CLC) are designed by using a mix design tool based on an optimized packing methodology. Applying the optimized packing method, the particles can be better packed, which results in improved hardened properties as well as improved workability, since more water is available to act as lubricant between the particles [15–18]. In this mix design method, the modified Andreasen and Andersen (A&A) model:

$$P(D) = \frac{D^q - D^q_{\text{min}}}{D^q_{\text{max}} - D^q_{\text{min}}} \tag{1}$$

where P(D) is a fraction of the total solids being smaller than size D, D is the particle size (μm), D_{max} is the maximum particle size (μm), D_{min} is the minimum particle size (μm) and q is the distribution modulus.

It acts as a target function for the subsequent granular optimization of the individual materials (detailed information is presented in [19–21]). The proportions of the individual materials in the mix are adjusted until an optimum fit between the composed mix grading curve and the target curve is reached, using an optimization algorithm based on the Least Squares Method (LSM), i.e. the deviation between the target curve and the composed mix expressed by the sum of the squares of the residuals (RSS) at defined particle sizes is minimal. Hence, the optimized mixture will possess a compact structure/matrix due to the optimal packing but also large value of non-interconnected pores, contributed by the LWA, which theoretically will lead to sufficient mechanical properties as well as good thermal insulation [3]. Table 1 shows the designed mixes and Fig. 1 shows the particle size distribution of the applied powders as well as the composed mixes.

Table 1 Characteristics of the designed concrete recipes

Material	Mix design (kg/m^3)
CEM III/A 52.5 N	429
Fly ash	45
Limestone powder	59
LWA 0.09-0.3	119.2
LWA 0.5-1	74.2
LWA 1-2	83.4
LWA 2-4	96.7
water	171.6
SP (% mass of cement)	1
Water/cement ratio	0.4
LDH (% volume of LWC)	1

2.3 Experimental Test

The experiments were performed on both fresh and hardened concretes. The flow table tests were performed following EN 12350-5:2009 [21]. The fresh density of the fresh concrete was determined following EN 12350-6:2009 [22], using a container with a volume of 10 dm^3. After preforming the workability test, the LWC was cast in moulds with the size of 40 mm × 40 mm × 160 mm for strength test and 160 mm × 160 mm × 160 mm for the RCM test. The prisms were demolded approximately 24 h after casting and then cured in water at about 21 °C. After curing for 1, 7 and 28 days, the flexural and compressive strengths of the specimens were tested

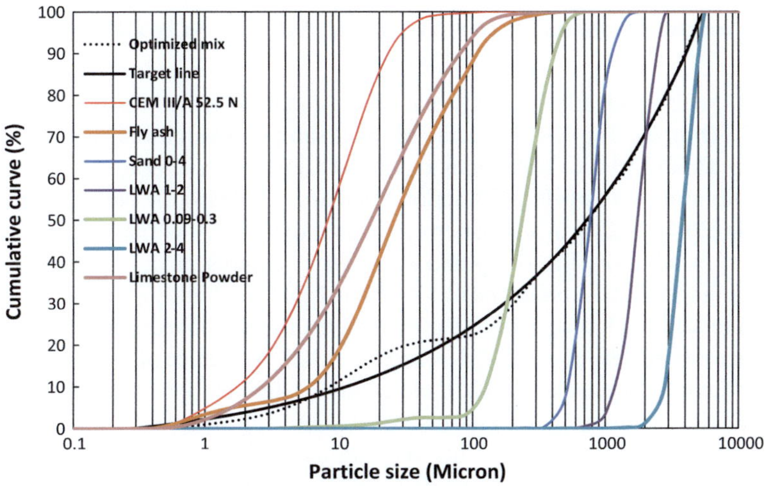

Fig. 1 Particle size distribution of the used materials and the composed mix

according to the EN 196-1 [23]. For the RCM test, cylinders were firstly drilled from cubes and cut to the required thicknesses. Cylindrical samples of 100 mm diameter and 50 mm thickness were used in the tests. The test samples were saturated with limewater under vacuum conditions, following the procedure specified in NT built 492 [24]. A pressure of about 20 mbar was applied to the samples in the desiccator for three hours, and then, with the vacuum pump still running, the desiccator was filled with limewater. The vacuum was maintained for an additional hour before allowing air to slowly re-enter the desiccator. The specimens were stored in the solution for 18 ± 2 h prior to the RCM test. Subsequently, the samples were placed in tightly clamped rubber sleeves. The catholyte was a 10% NaCl solution and the anolyte 0.3 M NaOH solution. The test duration and applied voltages were determined based on the initial current measured at 30 V. The RCM test duration for all the investigated mortar mixes was 24 h and the applied voltages are shown in Table 4. The chloride non-steady state migration coefficient (D_{nssm}) was calculated using Eq. (2).

$$D_{nssm} = \frac{0.0239(273 + T) \cdot L}{(U - 2) \cdot t} \cdot (x_d - 0.0238\sqrt{\frac{(273 + T) \cdot L \cdot x_d}{U - 2}}) \qquad (2)$$

where D_{nssm} is the non-steady state migration coefficient (10^{-12} m^2/s), U is the absolute value of the applied voltage (V), T is the average value of the initial and final temperatures in the solution (°C), L is the thickness of the specimen (mm), x_d is average value of the average chloride penetration depth (mm) and t is the test duration (h). Three specimen was tested for each concrete mixture at the age of 28 days.

3 Results and Discussion

3.1 *Characterization of the Synthetic Ca–Al–NO_3 LDHs*

The XRD analysis clearly confirms that LDHs can be obtained for at a pH value of 11 and 13 (Fig. 2). The XRD patterns of the 2 samples exhibit the characteristic [0 0 2], [0 0 4], and [0 0 6] reflections of Ca–Al LDHs with interlayer NO_{-3} anions (Joint Committee on Powder Diffraction Standards (JCPDS) file No. 89-6723). The first peak ($2\theta = 10.26$) in the pattern indicates that NO_{-3} is intercalated into the interlayer. In addition, the sharp and symmetric features of the diffraction peaks strongly suggest that the produced Ca–Al–NO_3 LDH was highly crystallized, having a three-dimensional lattice. All the samples have the rhombohedral space group $R\bar{3}c$ with lattice parameters of a = 5.731 Å and c = 48.32 Å. A trace of calcite is observed in the low angle range near 30° (2θ), indicating a slight carbonation of the samples.

The FTIR spectroscopy of Ca–Al–NO_3 LDHs with different sizes is shown in Fig. 3. The synthetic LDHs presents very similar FTIR spectroscopy pattern. The overlapping bands at 3483 cm^{-1} and 3636 cm^{-1} are attributed to the stretching vibrations of lattice water and OH^- associated to Ca^{2+} in Ca–Al–NO_3 LDHs, respectively

Fig. 2 XRD pattern of the Ca–Al–NO_3 LDH phases versus pH

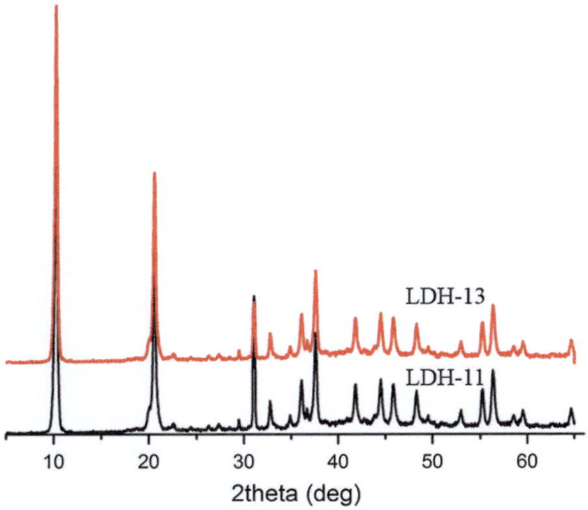

Fig. 3 The 4000–400 cm^{-1} region of FTIR spectra of the synthetic Ca–Al–NO_3 LDHs

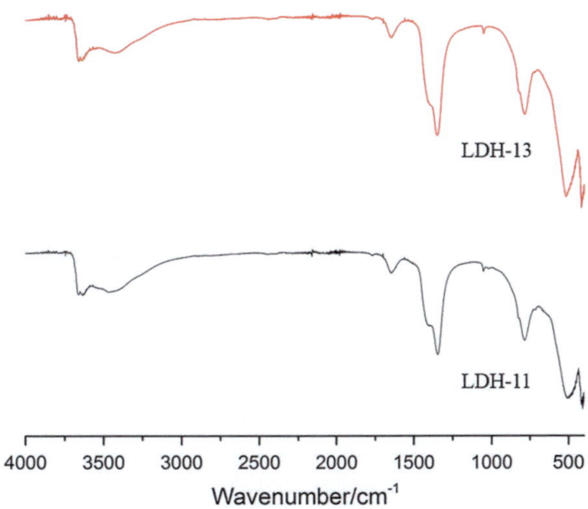

[25]. The peak at 1621 cm^{-1} shows the H–O–H bending vibration of the adsorbed water molecule. The peaks at 788 and 528 cm^{-1} reflect the stretching and deformation vibrations of M–OH. The anti-symmetric stretching vibration of NO_{-3} source is reflected by sharp split peaks at 1384 and 1344 cm^{-1} [13].

Figure 4 shows the morphology of Ca–Al–NO_3 LDHs prepared under pH = 11 and pH = 13. Under alkali condition, both of the prepared LDHs present a hexagonal flake crystal structure. With the increase of the pH value, the particle size of the LDHs decreases has shown in Fig. 4 and Table 2. The LDH-13 presents an average particle size of 2.2 μm in diameter while the LDH-11 presents an average particle size of 10.1 μm in diameter. This is in consistent with the specific area data. As

Fig. 4 SEM micrographs of three types of LDHs after sonication **a** pH = 11 **b** pH = 13

known, the decrease of the particles will result in higher specific areas. Hence, the LDH-13 exhibits a specific area of 50.50 m^2/g which is higher than that of LDH-11 (28.43 m^2/g). The differences in the crystallite sizes are obviously related to the crystallization pathway during the LDH formation. As we know, during the LDH formation and development reactions, precipitation of Al^{3+} and Ca^{2+} and dissociation of $Al(OH)_3$ and $Ca(OH)_2$ occur on the surface of the solid particles. So the speed of nucleation plays a key role in determination the growth of LDH crystals. It has been reported that Al^{3+} is first precipitating as a hydroxide [26]. Moreover, Ca^{2+} concentration is mainly decreasing as soon as there is no more aluminium in the solution. A higher pH value accelerates the growth of the LDH crystals, resulting in less time for crystallization and consequently smaller particle sizes.

Table 2 Physical property of the synthetic Ca–Al-NO_3 LDH

LDH	Diameter D (μm)	Thickness t (nm)	Aspect ratio, λ	Density g/cm^3	Specific Surface Area m^2/g
LDH-11	10.1	105	96	2.24	28.43
LDH-13	2.2	24	92	2.28	50.50

Table 3 Basic properties of the LWC

Mixture	Flow (cm)	Density (kg/m^3)	Oven dry density (kg/m^3)
Ref	65.5	1367	1245
LWC-LDH11	63.5	1378	1251
LWC-LDH13	62.5	1369	1249
LWC-LDH Mixed	61.5	1360	1248

3.2 *Fresh State of the LWC*

The spread flow of the mixture and the fresh density were measured and the results are shown in Table 3. The flowablity reduces slightly from the reference mixture, namely 65.5–63.5 cm (LWC-LDH11), 62.5 cm (LWC-LDH13) and 61.5 cm (LWC-LDH Mixed). It is clear that finer particles demand more water due to the higher specific surface area and consequently a lower flowability is found if the water dosage is fixed. Nevertheless, all the developed mixes show very good workability in overall, no traces of segregation or bleeding were observed while performing the workability tests. The good workability will decrease the generation of air pores in the LWC which is beneficial for the barrier effect of the LDH nano-flakes. The fresh densities of the mixtures are calculated from the measured mass and the fixed volume, yielding 1367, 1378, 1369 and 1360 kg/m^3 for the LWC-Ref, LWC-LDH11, LWC-LDH13 and LWC-LDH Mixed. It can be seen that the addition of LDH does not influence the fresh density of the LWC.

3.3 *Hardened State Properties of the LWC*

The oven-dry densities of the LWC are shown in Table 3 which are 1245, 1251, 1249 and 1248 kg/m^3 for the LWC-Ref, LWC-LDH11, LWC-LDH13 and LWC-LDH Mixed, respectively. This is consistent with the fresh density that the LDH nano-flakes do not influence the density of the LWC. The flexural and compressive strengths of the hybrid LWC at 1, 7 and 28 days versus the sizes of LDH are shown in Figs. 5 and 6 respectively. With the addition of LDH, an obvious increase tendency of the flexural strength of all the LWC can be observed. For example, the flexural

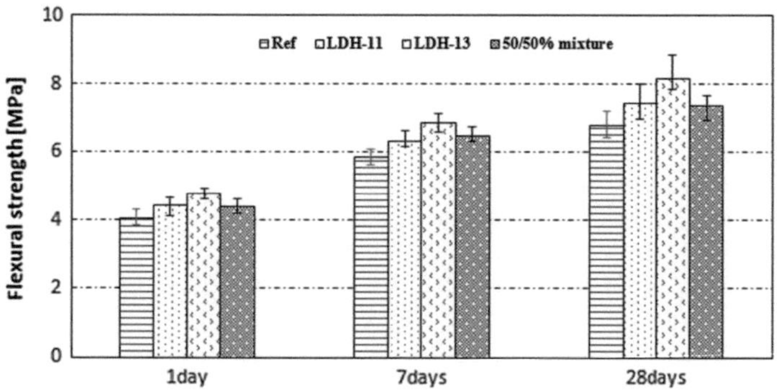

Fig. 5 Flexural strength of the developed LWC

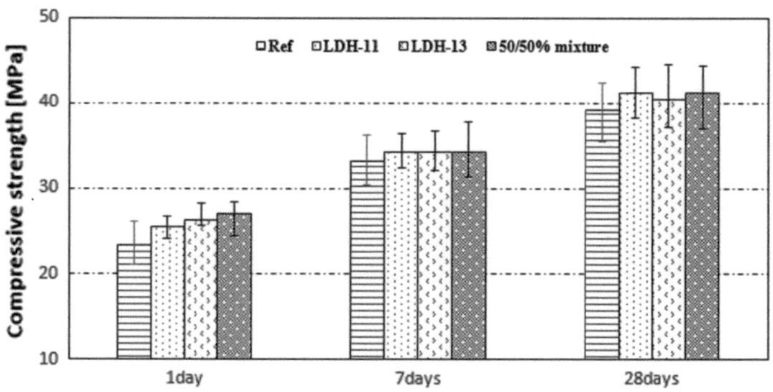

Fig. 6 Compressive strength of the developed LWC

strength of reference sample at 28 days is about 6.78 MPa, which gradually increases to about 7.44, 8.17 and 7.35 MPa separately when 1% (by volume) of LDH-11, LDH-13 and LDH-mixture were added respectively. This is attributed to the arrangement of nano-platelets which play a role as micro beam and form a stronger and stiffer mortar matrix. LDH-13 presents the biggest contribution on the flexural strength of the hybrid LWC. Comparing with the reference mixture, the LDH-13 increased the flexural strength by 18, 17 and 21% on 1, 7, 28 days respectively. This is due to the enhanced micro filler effect, resulting in a better interconnection of the CSH gel for the smaller size of the LDH-13.

The compressive strength of all the hybrid LWC improve slightly which also can be attributed to the micro filler effect of the cement paste. An addition of a crystal seed material with nanoscale size in the cement hydration process lowers the energy barrier to formation of the hydration products precipitated from the liquid, thus accelerating their rate of precipitation. The influence of the 2-D nano-fillers for the concrete mechanical properties is different from that reported in [9, 10]. This may be ascribed by the different physical states of the LDHs and GNP. Graphite type materials will bend under external force and this unique property has been used to prepare wearable electronic equipment and flexible screen [27]. Different from the flexible GNP, LDH is a kind of stiff enhanced unit element and it will transfer the stress efficiently and in turn improve the mechanical properties. In [19], stiff 2-D nano-fillers have been suggested to be used to prepare ultra-high performance concrete (UHPC) due to the excellent enhanced mechanical properties.

3.4 RCM Results of the LWC

The rapid chloride migration (RCM) test developed is one of the accelerated test methods in which chlorides penetrate the concrete at high rates due to the applied electric field [28]. The output of the test is the chloride diffusion coefficient D_{RCM} (often called the migration coefficient- to distinguish it from immersion tests). This method is concluded by some researchers to be the most suitable of all the reviewed accelerated chloride tests, on the basis of its simplicity, short duration and often is assumed to have a clear theoretical basis [28]. In the RCM test, after a period of the application of electric field to concrete, the chloride penetration depth in a split concrete sample is measured by an easy and quick $AgNO_3$ colorimetric method (0.1 mol dm^{-3} $AgNO_3$ solution sprayed onto the fractured concrete surface). The $AgNO_3$ colorimetric method involves two parameters–the indicated chloride penetration depth x_d and the free-chloride concentration c_d at which the color change occurs in concrete. Both these values are used to calculate the non-steady-state diffusion coefficient of chloride (D_{RCM}).

As stated in NT built 492, the free chloride penetration depth x_d is used for the calculation of the D_{RCM} according to Eq. (2) and it is determined by using the $AgNO_3$ solution. Theoretically, for a chloride contaminated concrete, only the chloride in pore solution can react with silver ions and yield the white precipitation. The study of Spiesz and Brouwers [28] has shown that the D_{RCM} is not affected by chloride binding. It is attributed to the fact that $AgNO_3$ colorimetric method was found reliable for the detection of the free chloride penetration front in the concrete. It has been demonstrated that the chloride binding capacity is very low at low free-chloride concentrations during the migration test and hence, the progress of the free chloride penetration front through the concrete sample is not retarded by chloride binding. Therefore, since in the migration tests chloride binding can be neglected at very low free chloride concentrations, the D_{RCM} calculated from the detected free chloride penetration front remains unaffected in the LDH contained lightweight concretes. Thus, the D_{RCM} can be used to characterize the physical barrier property of the LDH nano-flakes.

Tortuosity is a parameter describing an average elongation of fluid streamlines in a porous medium as compared to a free flow [30–32]. The introduction of the impermeable nano-flakes with high aspect ratio will lead to more tortuous paths for the ingress of fluids to the matrix materials. The tortuosity of penetration path for diffusing molecules is principally influenced by the following factors: the volume fraction of the nano-flakes; their morphologies (i.e., dispersion and orientation perpendicular to the diffusion direction) and their aspect ratio [33, 34]. As stated the in [11, 12], the apparent chloride diffusion coefficient D which can be expressed as:

$$D = \frac{D_{nssm}}{\tau^2} \tag{3}$$

where D is the apparent chloride diffusion coefficient, τ is the modified tortuosity factor and can be expressed as:

$$\tau = 1 + \frac{\lambda}{2}\phi \tag{4}$$

where λ and ϕ are the aspect ratio and the volume fraction of LDH in the matrix, respectively.

The average chloride penetration depth x_d is determined and used for the calculation of D_{RCM} Eq. (2), as reported in Table 4. The literature shows that the initial current value measured in the RCM test (at 30 V) may be correlated with the D_{RCM} values [29]. The initial current depends mainly on the chemistry of pore solution and pore structure of concrete [29]. The initial current measured at 30 V was 68.4 mA for reference, 38.7 mA for LWC-LDH11, 35.6 mA for LWC-LDH13 and 30.6 mA for LDH-LDH-Mixed which shows the same decrease trend as the D_{RCM}. It should be noted that the chloride diffusion coefficients for all the hybrid LWC are reduced compared to the reference, and this can be attributed to the barrier effect of the LDH nano-flakes which increase the total tortuosity of the LWC. Thus, the permeability of these LWC samples is greatly reduced and the chloride ingress resistance is improved. With as little as 1% addition amount, the D_{RCM} drops 25, 39, and 54% compared to the reference mixture. The resistance of the LWC to chloride migration with LDH is higher than that reported in literature for LWC containing nano-sized spherical particles nano-SiO_2 [10]. Similar to the mechanical properties, LDH-13 also performed better than the LDH-11 for the barrier effect. This can also be attributed to the enhanced micro filler effect of the smaller particle size of LDH-13. However, the 50–50% mixed LDH, whose average size is larger than LDH-13, presents the highest barrier effect to the LWC. This is attributed to the improved particle packing of the LWC system, the smaller LDH will fill in the space among the bigger ones, which in turn increases the possibility that the fluids encountering with the barrier and increases the average distance traveled by the fluids.

4 Conclusions

This research presents an investigation on the mechanical and transport properties of LWC incorporated with LDH nano-flakes, prepared with a simple co-precipitation method. Both the influence of the pH value on the sizes of LDH and the influence

Table 4 RCM test conditions and results

LWC Mixture	Sample	Applied Voltage (V)	Initial current (mA)	Test duration (h)	X_d	D_{RCM}
Ref	1	25	68.4	24	18.1	9.86
	2	–	–	–	9.45	9.91
	3	–	–	–	17.4	9.45
LWC-LDH11	1	30	38.7	24	16.3	7.35
	2	–	–	–	15.8	7.10
	3	–	–	–	16.5	7.44
LWC-LDH13	1	30	30.6	24	12.4	5.46
	2	–	–	–	13.6	6.04
	3	–	–	–	14.5	6.47
LWC-LDH Mixed	1	30	35.6	24	10.4	4.50
	2	–	–	–	9.8	4.22
	3	–	–	–	11.3	4.93

of nano-flakes sizes on the properties of the LWC were investigated. Based on the presented results, the following conclusions can be drawn:

- Through the control of pH values employing a co-precipitation route, LDHs with different sizes are prepared. The XRD and FTIR analyses confirm that the as-prepared LDHs possess the same phases and function groups.
- The designed reference lightweight concrete shows excellent mechanical properties, i.e. a 28-day compressive strength of about 40 MPa with an oven dry density of 1245 kg/m^3.
- The size of the nano-flakes plays a great role on the mechanical properties of the lightweight concrete due to the micro-filler effect. The smaller sizes nano-flakes performance better in both mechanical and transport properties, e.g. the LDH-13 can increase the flexural strength with 20%.
- The addition of LDH nano-flakes significantly improves the resistance of concrete to chloride ion penetration due to the increase of tortuosity. The resistance of the LWC to chloride migration increases up to 54% with the addition of LDH.

References

1. Yu QL, Spiesz P, Brouwers HJH (2015) Cem Concr Compos 61:18–28
2. Zhang MH, Gjvorv OE (1991) Mater J 88(3):240–247
3. Yu QL, Spiesz P, Brouwers HJH (2013) Cem Concr Compos 44:17–29
4. Spiesz P, Yu QL, Brouwers HJH (2013) Cem Concr Compos 44:30–40

5. Quercia G, Spiesz P, Hüsken G, Brouwers HJH (2014) Cem Concr Compos 45:69–81
6. Tang SW, Yao Y, Andrade C, Li ZJ (2015) Cement Concrete Res 78:143–154
7. Gao X, Yu QL, Brouwers HJH (2015) Constr Building Mater 98:397–406
8. Quercia G, van der Putten JJG, Hüsken G, Brouwers HJH (2013) Cement Concrete Res 54:161–179
9. Sanchez F, Sobolev K (2010) Constr Building Mater 24(11):2060–2071
10. Zhang MH, Gjørv OE (1992) Cement Concrete Res 22(1):47–55
11. Du H, Dai Pang S (2015) Cement Concrete Res 76:10–19
12. Du H, Gao HJ, Dai Pang S (2016) Cement Concrete Res 83:114–123
13. Yang Z, Fischer H, Polder R (2014) Cem Concr Compos 47:87–93
14. Chen Y, Shui Z, Chen W, Chen G (2015) Constr Building Mater 93:1051–1058
15. Quercia Bianchi G, Hüsken G, Brouwers HJH (2012) Cem Concr Res 42:344–357. Erratum, ibid, 46, 73–74 (2013)
16. Yu QL, Brouwers HJH (2011) Constr Building Mater 25:3149–3157
17. Hunger M, Brouwers HJH (2009) Cem Concr Compos 31:39–59
18. Brouwers HJH (2006) Phys Rev E 74(3):031309
19. Yu QL, Brouwers HJH (2012) Cem Concr Compos 34:1033–1043
20. Hüsken G, BrouwersHJH (2008) Cem Concr Res 38:1246–1259. Erratum, ibid 39, 832 (2009)
21. EN, B. (2000) 12350-5. Testing fresh concrete-Part-5: flow table test. British Standard
22. EN B (2009) 12350-6. Testing Fresh Concrete Density. British Standards Institution
23. EN T (2005) 196-1. Methods of testing cement–Part 1: determination of strength. European Committee for standardization, 26
24. Built NT 492 (1999) Chloride migration coefficient from non-steady-state migration experiments. Nordic council minister
25. Chen Y, Shui Z, Chen W, Chen G (2015) Constr Building Mater 93:1051–1058
26. Seron A, Delorme F (2008) J Phys Chem Solids 69(5):1088–1090
27. Liao X, Liao Q, Yan X, Liang Q, Si H, Li M, Zhang Y (2015) Adv Functional Mater 25(16):2395–2401
28. Carnelli D, Libanori R, Feichtenschlager B, Nicoleau L, Albrecht G, Studart AR (2015) Cement Concrete Res 78:245–251
29. Spiesz P, Brouwers HJH (2013) Cement Concrete Res 48:116–127
30. Maxwell JC (1881) Treatise on Electricity and Magnetism. Clarendon Press, Oxford
31. Moggridge GD, Lape NK, Yang C, Cussler EL (2003) Prog Organic Coatings 46(4):231–240
32. DeRocher JP, Gettelfinger BT, Wang J, Nuxoll EE, Cussler EL (2005) J Membrane Sci 254(1):21–30
33. Cussler EL, Hughes SE, Ward WJ, Aris R (1988) J Membrane Sci 38(2):161–174
34. Cussler EL (2009) Diffusion: mass transfer in fluid systems. Cambridge university press

Using Thymolphthalein for Accelerated Carbonation Testing of High Volume Fly Ash Cementitious Blends

Rui Reis, Aires Camões and Manuel Ribeiro

Abstract Phenolphthalein is the most common pH indicator used to measure the carbonation depth of cementitious materials. It is simple and quick to use, but according to some authors, is not an appropriate tool. Moreover, this substance is in the candidate list as a substance of very high concern by European Chemicals Agency. This article discusses the use of thymolphthalein rather than phenolphthalein. Accelerated carbonation tests were performed on cement mortars and high volume fly ash mortars. Once this kind of mixtures has low carbonation resistance additional alkalinity was provided by adding hydrated lime during its fresh mixture, and by curing the samples immersed in a calcium hydroxide solution. The results showed that the best indicator was the thymolphthalein, mainly for blends with lower carbonation depth. It was also checked that adding alkalinity can be useful for increasing the carbonation resistance.

1 Introduction

The most widely used method to evaluate the carbonation front of concrete uses phenolphthalein as a pH indicator [16, 21–23], but some authors point out weaknesses in the use of this methodology. According to Houst and Wittmann [8] the carbonation depth measured with phenolphthalein does not match the carbonation profile. In addition, the carbonation depth spread mechanism is complex and cannot be described by a simple method. According to Torgal et al. [21] phenolphthalein is a poor indicator to estimate the carbonation depth and their widespread use in the investigation of carbonation may have hindered some progress in the acquired

R. Reis (✉) · A. Camões
CTAC, Department of Civil Engineering, University of Minho, Campus de Azurém, 4800-058 Guimarães, Portugal
e-mail: ruicunhareis@gmail.com

M. Ribeiro
UIDM, ESTG, Polytechnic Institute of Viana do Castelo, 4900-347 Viana do Castelo, Portugal

C. Andrade et al. (eds.), *Service Life and Durability of Reinforced Concrete Structures*, RILEM Bookseries 17, https://doi.org/10.1007/978-3-319-90236-4_2

knowledge. For Khunthongkeaw et al. [10], the beginning of reinforcement corrosion is considered as one of the critical stages for predicting service life of a concrete structure. Therefore, it is necessary to estimate the precise time when the carbonation front reaches the reinforcing steel bars. Thus, if the phenolphthalein methodology is not sufficiently precise, it may be convenient to use a more accurate tool to anticipate the results.

The advantage of pH indicators is that they allow a simply and quickly evaluation, a boundary where one side is healthy and the other is carbonated. Nevertheless, this division is not perfectly clear, since the indicators only differentiate fully carbonated from uncarbonated concrete. Therefore, with this methodology the carbonation is not considered gradual, but abrupt. However, carbonation penetrates into concrete [8, 11], a first zone near the surface, completely carbonated or homogeneously carbonated; a second zone, of transition, called carbonation front, where the degree of carbonation decreases to zero; and a third zone with no trace of carbonation. According to Siddique [19], it is possible to relate the pH of the pore solution with the porous degree of carbonation, but there are few studies on this topic [14].

The carbonation front may be evaluated indirectly and roughly by an acid-base indicator, through pH [12]. Various indicators are available, such as phenolphthalein, thymolphthalein, cresol red, or alizarin yellow [20]. For instance, phenolphthalein has a pH range between 8.2 and 9.8 (colourless to purple) [12], and thymolphthalein between 9.3 and 10.5 (colourless to blue) [16]. The use of different indicators concentrations must also be considered. In order to compare the obtained test results the pH indicators concentrations should be uniform in all investigations, which does not happen [6, 12, 16]. Perhaps because of this, some authors suggest a different pH range: according to Cunha and Helene [7], phenolphthalein has a pH range between 8.2 and 9.8; according to Younsi [23], the turning takes place around pH 9.

In addition to the above mentioned and according to FprCEN/TS 12390-12 [6], in 2011, phenolphthalein was included in the ECHA candidate list as a substance of very high concern, due to the following risks: R45 carcinogenic—category 2; R68 mutagenic—category 3; R62 toxic to reproduction—category 3. Therefore, this document recommend thymolphthalein rather than phenolphthalein. Thus, differences between these two acid-base indicators must be evaluated.

In this study were proposed cementitious mixtures with high volume of fly ash, because the alkali compounds, in particular the calcium hydroxide (CH), reacts with fly ash, and thus decreases its pH. Therefore, it is expected that such mixtures have reduced carbonation resistance [23]. However, the importance of curing (manly wet curing by immersion) increases with the amount of incorporated mineral additions, namely pozzolanic ones [5, 17]. High volume fly ash concrete associated with a dry curing process may have extremely severe carbonation [13], thus it was intended to carry out wet curing by water immersion, and also wet calcium hydroxide curing by immersion into CH saturated solution. In fact, the curing process has a great influence in concrete durability and carbonation speed. A wet curing promotes a greater compounds hydration (mainly on the outside of the concrete), resulting a lower porosity and more compact concrete that decreases the diffusion of CO_2 [2], and so, turns it less susceptible to carbonation [23].

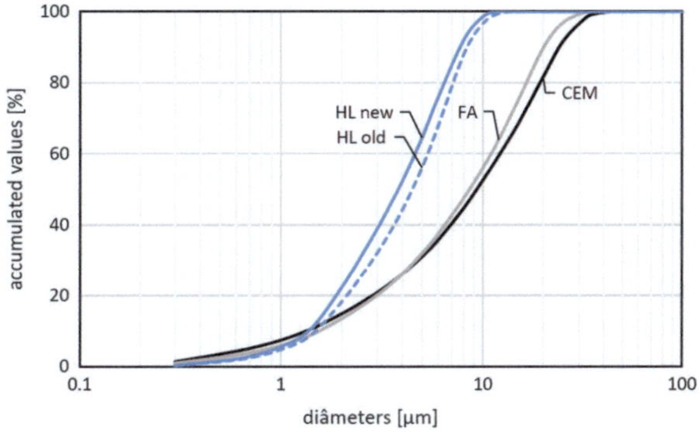

Fig. 1 Particle size analysis

2 Experimental Program

The selected materials used during the experimental campaign were: cement CEM I 42.5R; river rolled sand 0/4 mm; fly ash from the Portuguese Pego thermoelectric power plant; calcium hydroxide in the form of hydrated lime; and a superplasticizer based on modified polycarboxylic. The particle size of these powders was carried out by laser analysis which results are presented in Fig. 1. It was found that the two tested binders had similar fineness (cement and fly ash), although the fly ash had a slightly lower fineness, except for the smaller particles. In the case of hydrated lime, it was considered important to evaluate the product stored in two different conditions: in a closed bag (as used in this study—HL New); in an open bag—HL Old. With this procedure one intends to evaluate the importance of using uncarbonated hydrated lime or partially air carbonated hydrated lime. Analysing the particle size results it was verified that hydrated lime is finer than the selected cement and fly ash. There was also a slightly difference between the closed and open hydrated lime bag, which should be carefully considered when working with this kind of material.

Simultaneous thermal analysis (*STA*), at *Netzsch* 402 *EP* equipment, heating rate of 10 °C/min, were also carried out for the hydrated lime. Figure 2 shows the *STA* curves for the closed bag lime and open bag lime. Big differences in the peaks relating to water (*A*), portlandite or CH (*B*) and calcite or $CaCO_3$ (*C*) were observed. The reason for this difference may be easily explained: when the bag is open, calcium hydroxide reacts with the atmosphere and produces calcium carbonate. Thus, this research was performed with sealed bags, considering a CH content of 75.8% and $CaCO_3$ content of 22.6%. However, it was clear shown that if there is no extreme caution, it would be possible that CH content drops to 58.2% and $CaCO_3$ can rise to 40.7%.

Fly ash was characterized with a trial of X-ray diffraction (*XRD*) and a trial X–ray fluorescence spectroscopy (*XRF*), including the identification and quantification of reactive phase by Rietveld method. This evaluation was conducted in a previous study [15] and the results are presented in Table 1. A scanning electron microscope

Hitachi SU1510 (*SEM*) was used for a qualitative evaluation. It was found that: fly ash consists mainly of cenospheres particles (Fig. 3–1); a small amount is trapped in carbon clusters (Fig. 3–2); were almost not found impurity, just quartz, mullite (Fig. 3–1); and hematite (3–4).

The research was developed based on a previously studied [4] plain self-compacting high performance concrete, made with 500 kg/m^3 of cement (*C*), 731 kg/m^3 of aggregate (*S*) and water-binder ratio (*W/B*) of 0.25. Mortar samples were produced by keeping the previous proportion, for 3 water-binder ratio (0.25, 0.30 and 0.35) and 3 distinct groups: plain cement (*C*); cement replaced by 50% of fly ash (*FA*) by weight of binder ($B = C + FA$); cement replaced by 50% fly ash by weight of binder, plus hidrated lime (*HL*) as an aggregate replacement. It was used a constant ratio of superplasticizer (*SP*), 1.70% by weight of the binder. The summary of mixtures is shown in Table 2.

In Fig. 4 one can see the adopted experimental work schedule. For each mixture were produced 2 samples with $40 \times 40 \times 160$ mm^3. After 24 h, the samples were demoulded [9], and subjected to wet curing (immersed in water) until 38 days of age. After, one of the samples remains in wet curing (*H*) and the other in water saturated with calcium hydroxide curing (*CH*). After, the samples were subjected at 40 °C for 2 days in a greenhouse. Then, all the sample's surfaces except two opposite ones were sealed with paraffin and subsequently stored in a carbonation chamber with $4 \pm 0.5\%$ CO_2, $55 \pm 5\%$ RH and 20 ± 2 °C [6]. Later after 30 days, two cross-sections were taken off of each sample, to measure the carbonation depth, using thymolphthalein and phenolphthalein acid-base indicator. Finally, the samples were sealed again with paraffin, remaining in the chamber over more 70 days, for a total of 100 days of accelerated carbonation test.

To calculate the carbonation depth, each exposed face was divided in 6 equidistant sections [6]. The carbonation depths were then measured. It was considered as carbonation depth of each mixture, the arithmetic average of the both exposed

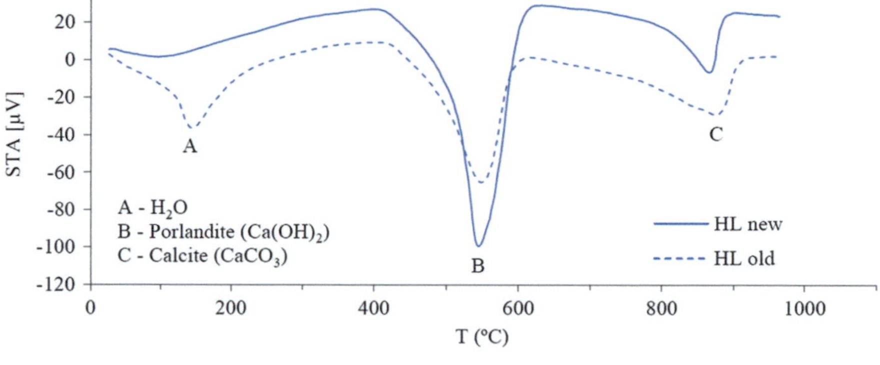

Fig. 2 STA hydrated lime

Table 1 Materials characteristics

	C I 42.5R	FA Pego			HL Calcidrata
Chemical properties (%)	[18]	XRF	MEV	XRD	STA/XRD [3]
SiO_2	18.27	49.12	56.83	28.7 vitr.*	< 0.4 [3]
Al_2O_3	4.75	27.30	28.59	22.1 vitr.*	< 0.5 [3]
Fe_2O_3	3.23	8.19	6.86	4.8 vitr.*	< 0.08 [3]
CaO	63.94	2.36	1.80		
CaO free	1.39				
MgO	1.31	1.42			< 0.85 [3]
SO_3	3.05	1.30			
K_2O	0.57	3.34	1.97		
Na_2O		0.99	1.78		
TiO_2		2.32			
Cl^-	0.09				
Other				15.4 vitr.*	
Carbon			3.65		
Loss on ignition	2.72	3.97			
Insoluble residue	0.74				
Physical properties					
D10 (laser an.)	1.33		1.53		1.36
D50 (laser an.)	9.30		8.55		3.80
D90 (laser an.)	24.16		20.19		7.73
Sp. gravity (kg/m^3)	3100	2420			2240 [3]
Blaine [cm^2/g]	4315				
Mineralogical properties (%)					
Quartz				18.41	
Mullite				7.18	
Hematite				3.41	
Total crystalline phase*				29.00	
Total glassy phase				71.00	
Portlandite					75.8
Calcite					22.6

*vitreous phase by Rietveld method

Table 2 Tested compositions

Designation		W/B	Materials						
			B	C	FA	HL	S	W	SP
			(kg/m^3)	(kg/m^3)	(kg/m^3)	(kg/m^3)	(kg/m^3)	(kg/m^3)	(%L)
I	100 °C	0.25	856	856	0	0	1252	214	1.70
		0.30	856	856	0	0	1141	257	1.70
		0.35	856	856	0	0	1029	300	1.70
II	50C + 50FA	0.25	856	428	428	0	1151	214	1.70
		0.30	856	428	428	0	1040	257	1.70
		0.35	856	428	428	0	928	300	1.70
III	50C + 50FA + 2.5HL	0.25	856	428	428	21.4	1126	214	1.70
		0.30	856	428	428	21.4	1015	257	1.70
		0.35	856	428	428	21.4	903	300	1.70
IV	50C + 50FA + 5HL	0.25	856	428	428	42.8	1101	214	1.70
		0.30	856	428	428	42.8	990	257	1.70
		0.35	856	428	428	42.8	878	300	1.70
V	50C + 50FA + 10HL	0.25	856	428	428	85.6	1051	214	1.70
		0.30	856	428	428	85.6	940	257	1.70
		0.35	856	428	428	85.6	828	300	1.70

(continued)

Table 2 (continued)

Designation		W/B	Materials						
			B	C	FA	HL	S	W	SP
			(kg/m³)	(kg/m³)	(kg/m³)	(kg/m³)	(kg/m³)	(kg/m³)	(%L)
VI	50C + 50FA + 15HL	(0.25)	856	428	428	128	1002	214	1.70
		0.30	856	428	428	128	890	257	1.70
		0.35	856	428	428	128	779	300	1.70
VII	50C + 50FA + 20HL	(0.25)	856	428	428	171.2	952	214	1.70
		0.30	856	428	428	171.2	840	257	1.70
		0.35	856	428	428	171.2	728	300	1.70
VIII	50C + 50FA + 30HL	(0.25)	856	428	428	257	852	214	1.70
		0.30	856	428	428	257	740	257	1.70
		0.35	856	428	428	257	628	300	1.70

(W/B) was not possible to produce, due to its low workability

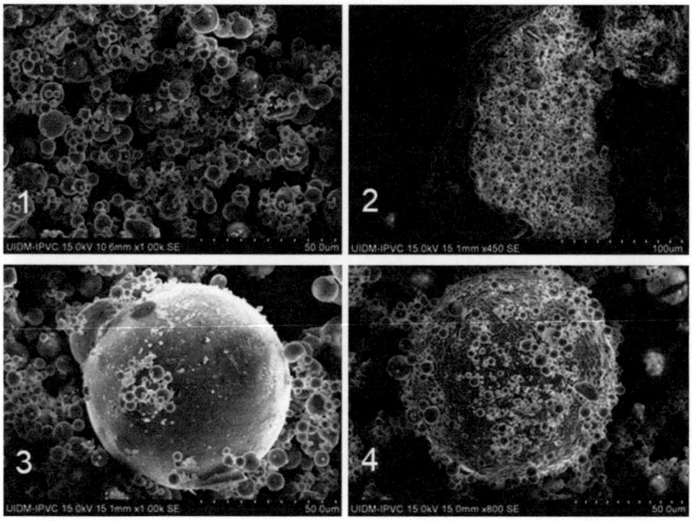

Fig. 3 SEM fly ash images

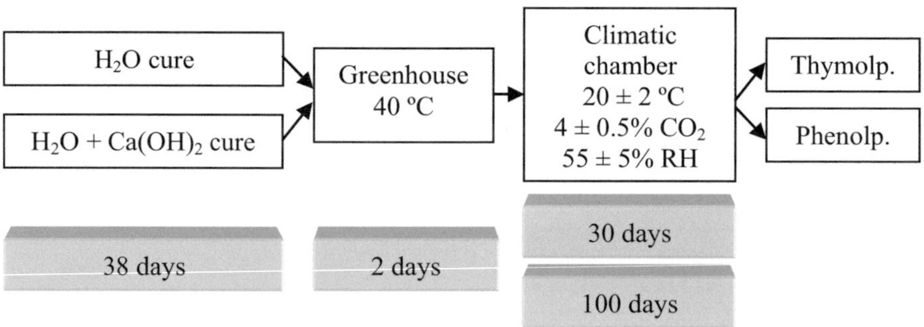

Fig. 4 Experimental work schedule

side's measured. In Fig. 5 is shown an example of measurement with two acid-base indicators.

3 Analysis of Results

3.1 Thymolphthalein and Phenolphthalein Indicator

Four tests of carbonation depth were carried out: 30 and 100 days after accelerated carbonation in the climatic chamber. Two acid-base indicators were used to measure the carbonation depth: thymolphthalein and phenolphthalein. Figure 6 shows the

Fig. 5 Samples on the climatic chamber (left). Phenolphthalein (center) and thymolphthalein (right) measurements

results of carbonation depth, after 100 days of exposure on the carbonation chamber, in wet curing (H) and in water saturated with calcium hydroxide curing (CH). Absolute results of these measurements show the same tendencies and can be considered as roughly similar ones. Furthermore, there was a very obvious difference: even after 100 days of accelerated carbonation test, phenolphthalein has not revealed visible signs of carbonation for mixtures with plain cement (mixture I). This fact can make a difference: It would be necessary to extend the maturity of the test on the carbonation chamber.

To show the relative differences between carbonation depth measurements with thymolphthalein and phenolphthalein, thymolphthalein- phenolphthalein depth ratio was calculated according to Eq. (1)

$$ratio = \frac{thymolphthalein_{depth} - phenolphthalein_{depth}}{thymolphthalein_{depth}} \quad (1)$$

Figure 7 shows the thymolphthalein- phenolphthalein depth ratio after 100 days climatic chamber. The biggest amount of values presented are positive ones, which means that the use of thymolphthalein provides higher values. This is in accordance with the state of the art cited above, because the thymolphthalein pH range is higher than phenolphthalein. It was also found that the samples subjected to wet calcium hydroxide curing, shown, in general, a greater ratio. This suggests that it might be caused by the higher resolution of thymolphthalein: the carbonation depth of this sample was lower, which highlight the highest easily measurement of thymolphthalein rather of phenolphthalein. Probably for the same reason, the mixtures IV and V (with less carbonation depth), had higher ratio.

3.2 *Wet Curing and Wet Calcium Hydroxide Curing*

Figure 8 shows the results of the water—calcium hydroxide depth curing ratio, after 100 days of exposure on carbonation chamber, calculated according to Eq. (2).

$$ratio = \frac{water_{depth} - calciumhydroxide_{depth}}{water_{depth}} \quad (2)$$

It's also shown the difference between the use of thymolphthalein and phenolphthalein. There was a slight tendency for the results to locate in the negative axis. This may mean that the use of saturated calcium hydroxide curing solution can provide greater alkalinity or the cure might have avoid leaching of alkaline compounds of the samples. It was also found that the measured values using thymolphthalein are generally higher than those measured with the phenolphthalein. The reasons are justified above.

3.3 *Effect of Time During the Accelerated Test*

To compare the carbonation depth with different maturities in the climatic chamber one has used the simplified model k-n usually used for natural carbonation [1], according to Eq. (3).

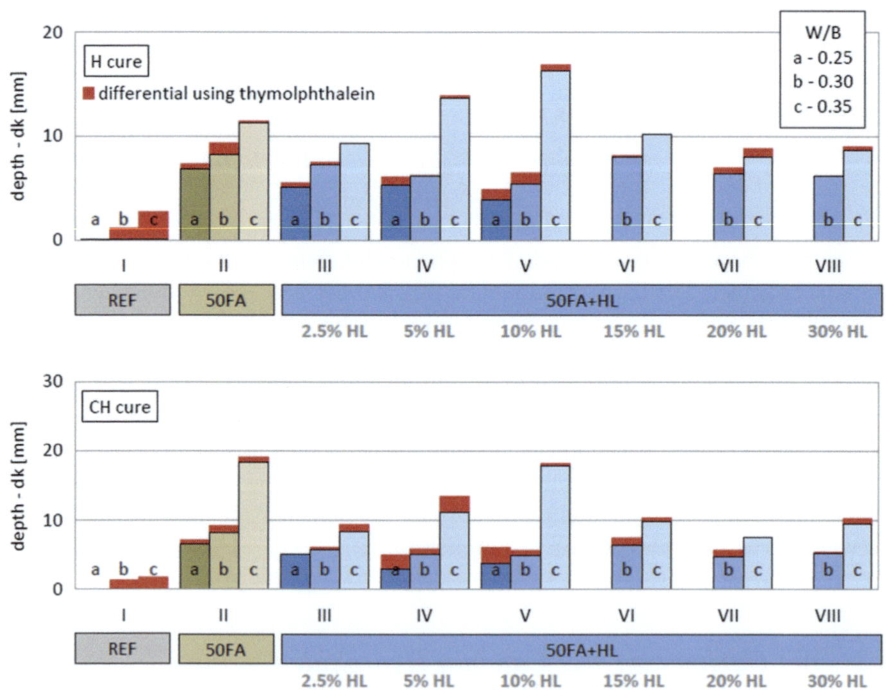

Fig. 6 100 days carbonation depth

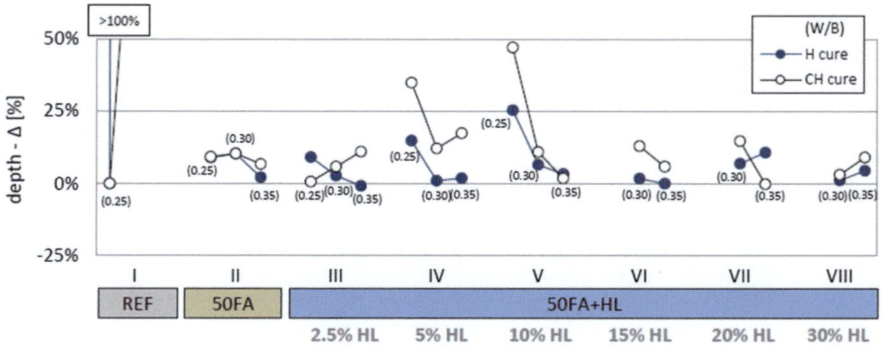

Fig. 7 100 days thymolphthalein—phenolphthalein depth ratio

$$x = K \cdot t^n \tag{3}$$

K is a carbonation constant (mm/yearn); t is the time in the climatic chamber (years); n is a factor which depends on the environmental conditions and the type of concrete, was considered equal to 5 (typical value). From the values of the carbonation depth (x) the K for accelerated test at 30 ($K30$) and 100 days ($K100$) were determined. Was calculated the ratio between $K100$ and $K30$ according to Eq. (4).

$$ratio = \frac{K100_{depth} - K30_{depth}}{K100_{depth}} \tag{4}$$

Figure 9 shows the $K100$–$K30$ depth ratio, by thymolphthalein, for the two kind of curing process. In a simplified way, it would be expected that the ratio was near to 1. It was found for the samples which were in wet calcium hydroxide curing, almost all showed a ratio equal or below 1. For the samples which were in wet curing, the results are very dependent of the composition. This may mean that the carbonation depth decreases with time. Also, the studied compositions represent materials with

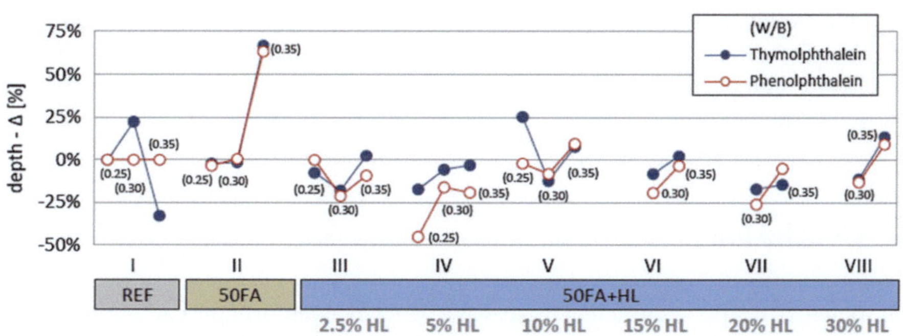

Fig. 8 100 days water—calcium hydroxide depth curing ratio

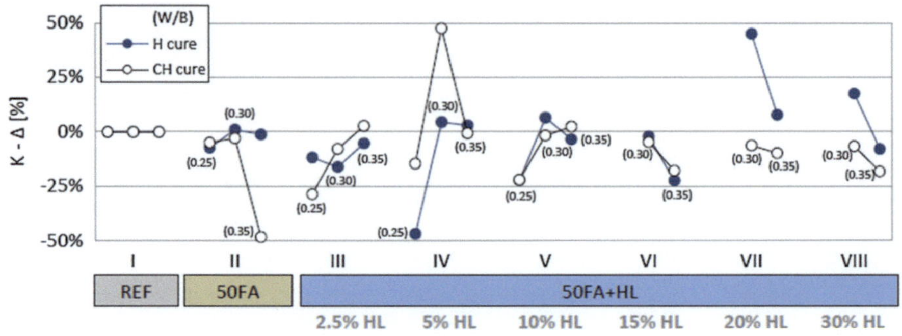

Fig. 9 K100—K30 depth ratio by thymolphthalein

different characteristics and behavior, so the power factor (n) of the Eq. (3) must be adjusted for each composition and kind of curing process [1]. On the other hand, it seems that the wet calcium hydroxide curing may uniform the determined ratios for the different tested mixtures.

3.4 *Carbonation Depth*

Figure 10 represents the carbonation depths, obtained through accelerated tests at 100 days of exposure on carbonation chamber and using thymolphthalein as pH indicator. First, it was found that the carbonation depth is greater for pozzolanic mixtures (II–VIII). This is probably due to the consumption of CH during the pozzolanic reaction. In this reaction, which describes the hydration of the fly ash, CH is required as a reagent. It was also found that, the use of hydrated lime was most advantageous from 2.5 to 10% incorporation (mixtures III–V). It was even possible to reduce the carbonation depth up to 57% (mixture V-*a* rather II-*a*) compared to mixtures containing just 50C + 50FA.

For hydrated lime content above 10%, the results were not as expected. One possible explanation could be the fact that the excessive increase of hydrated lime, does not find more glass phase of the fly ash available to react. It can also be justified by a lower compactness of the matrix, whereby the excess alkali is not enough to prevent carbonation.

4 Conclusion

The obtained results suggest that there was a better evaluation of the carbonation front when the thymolphthalein indicator was used. It was also found that water saturated with wet calcium hydroxide curing process is not, by itself, sufficient to reduce the

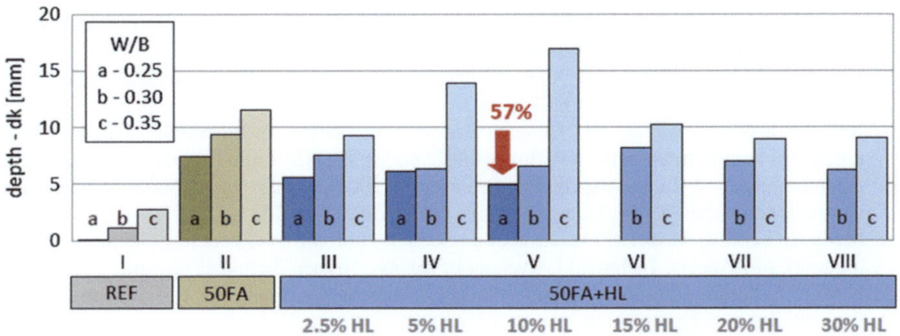

Fig. 10 100 days carbonation depth by thymolphthalein

carbonation. But, the addition of hydrated lime in mixtures with high fly ash content can be a tool to mitigate carbonation, typical of this kind of mixtures. Unlikely as happens in the cement hydration, calcium hydroxide is added at the beginning of the fresh mixture. Thus, a change may happen in the reaction kinetics and may also result in reaction products with different characteristics and behaviours. For these reasons further investigation is required.

Acknowledgments The authors wish to acknowledge the team of the Laboratory of Building Materials of University of Minho and the Unit for Research and Development Materials of ESTG from Polytechnic Institute of Viana do Castelo in this research collaboration. The authors also wish to thank Secil SA, BASF SA and Mota Engil by its kindly supply of cement, superplasticizer and fly ash, respectively.
SEM and XRD facilities were funded by FEDER Funds through QREN—Aviso SAIECT-IEC/2/2010, Operação NORTE-07-0162-FEDER-000050.

References

1. Alexander MG, Mackechnie JR, Yam W (2007) Carbonation of concrete bridge structures in three South African localities. Cement Concrete Compos 29(10):750–759
2. Atiş CD (2003) Accelerated carbonation and testing of concrete made with fly ash. Constr Building Mater 17(3):147–152
3. Calcidrata (2010) Product specification sheet: hydrated lime, class 1, grade A, type 1, according to EN 12518: 2008 (in Portuguese). Calcidrata-indústrias de cal SA
4. Camões A (1996) (pdf). Use of fly ash in concrete. In: V. Baroghel-Bouncy et al. (eds) International RILEM workshop on performance based evaluation and indicator for concrete durability. Madrid, Spain, 19–21 March, ACI Committee 232: ACI Committee 232
5. Casore W (2008) Study of the carbonation of the cover layer of concrete prototypes with high contents of mineral additions and hydrated lime (in Portuguese). MSc thesis. Federal University of Santa Maria
6. CEN (2010) Testing hardened concrete-Part 12: Determination of the potential carbonation resistance of concrete: Accelerated carbonation method FprCEN/TS 12390-12:2010. CEN (European Committee for Standardization), Brussels

7. Cunha A, Helene P (2001) Depassivation of concrete armor by carbonation (in Portuguese). Polytechnic School of the University of São Paulo, São Paulo
8. Houst YF, Wittmann FH (2002) Depth profiles of carbonates formed during natural carbonation. Cement Concrete Res 32(12):1923–1930
9. IPQ, 2011. NP EN 197-1:2011 Cement. Part 1: Composition, specifications and conformity criteria for common cements (in Portuguese). Caparica: IPQ (Instituto Português da Qualidade)
10. Khunthongkeaw J, Tangtermsirikul S, Leelawat T (2006) A study on carbonation depth prediction for fly ash concrete. Constr Building Mater 20(9):744–753
11. Monteiro I (2010) Model of evolution of carbonation in reinforced concrete elements (in Portuguese). MSc thesis. Technical University of Lisbon. Higher Technical Institute
12. Neville A (2003) Can we determine the age of cracks by measuring carbonation? Part 1. Concrete Int 25(12):76–79
13. Ngala VT, Page CL (1997) Effects of carbonation on pore structure and diffusional properties of hydrated cement pastes. Cement Concrete Res 27(7):995–1007
14. Nilsson L-O, Fridh K (2009) CO_2-cycle in cement and concrete: part 1: examination of a previous NIC-project - part 2: further literature study on mechanisms/models. [pdf] Lund: L. University
15. Reis R, Malheiro R, Camões A, Ribeiro M (2014) Carbonation resistance of high volume fly ash concrete. Key Eng Mater 634(2015):288–299
16. Rincón O, Aleida C, Carmen A, Paulo H, Isabel D (eds) (1997) Manual for inspection, evaluation and diagnosis of corrosion in reinforced concrete structures (in Spanish). CYTED (Programa Iberoamericano de Ciência y Tecnologia para el Desarrollo), Habana
17. Sasatani T;Torii K, Kawamura M (1995) Five-year exposure test on longterm properties of concretes containing fly ash, blast-furnace slag, and silica fume. In: 5th International conference on the use of fly ash, silica fume, slag, and natural pozzolans in concrete, Milwaukee, ACI SP-153
18. SECIL (2013) Bulletin N. ACO-001/2013—portland cement—EN 197-1 CEM I 42.5 R: results of Self-control (in Portuguese). Outão, SECIL
19. Siddique R (2011) Properties of self-compacting concrete containing class F fly ash. Mater Design 32(3):1501–1507
20. Thiery M (2005 Model of atmospheric carbonation of cementitious materials: taken into account kinetic effects and microstructural and water changes (in French). PhD thesis. Ecole des Ponts Paris Tech
21. Torgal FP, Miraldo S, Labrincha JA, De Brito J (2012) An overview on concrete carbonation in the context of eco-efficient construction: evaluation, use of SCMs and/or RAC. Constr Building Mater 36(2012):141–150
22. Villain G, Thiery M, Platret G (2007) Measurement methods of carbonation profiles in concrete: thermogravimetry, chemical analysis and gammadensimetry. Cement Concrete Res 37(8):1182–1192
23. Younsi A (2011) Carbonatation de bétons à forts taux de substitution du ciment par des additions minérales. PhD thesis. Université de la Rochelle

Durability and Mechanical Properties of CNT Cement Composites

Carmen Camacho-Ballesta, Óscar Galao, Francisco Javier Baeza, Emilio Zornoza and Pedro Garcés

Abstract Due to their unique intrinsic properties, carbon nanotubes (CNT) are considered potential candidates for developing new functional properties when they are included into the cementitious matrix. This work has the aim of characterising the main properties of CNT Portland cement composites, regarding their mechanical properties and their durability facing corrosion processes. Variation in mechanical properties of mortars with different dosages of CNT and corrosion of embedded steel rebars in CNT cement pastes were investigated. Firstly, bending strength, compression strength, porosity and density of CNT mortars were obtained and compared with the reference (without CNT). Afterwards, CNT reinforced paste specimens were prepared to be exposed to carbonation and chloride attacks. The results on steel corrosion rate tests were related to CNT dosages. The increase in CNT addition implied no significant variations of mechanical properties but slightly higher steel corrosion intensities were found.

1 Introduction

Today's demand for smart structures, capable of detecting stimulus and responding adequately, has created the need for materials with not only good mechanical properties and durability, but also new additional functions. That is the reason why many researches have been focused on the development of multifunctional materials, which combine both structural properties and other functional applications [1].

Multifunctional cement–matrix composites are useful as structural materials that provide other functional properties, which allow applications such as electrical grounding, electrical contacts for cathodic protection and electrochemical chloride extraction [2], deicing [3], electromagnetic interference (EMI) shielding [4], antistatic flooring and strain and damage sensing [5–7]. Multifunctionality is attractive for cost reduction, durability enhancement, large functional volume, design simpli-

C. Camacho-Ballesta (✉) · Ó. Galao · F. J. Baeza · E. Zornoza · P. Garcés
Civil Engineering Department, University of Alicante, Alicante, Spain
e-mail: mc.camacho@ua.es

C. Andrade et al. (eds.), *Service Life and Durability of Reinforced Concrete Structures*, RILEM Bookseries 17, https://doi.org/10.1007/978-3-319-90236-4_3

fication and absence of mechanical property loss (which tends to occur if embedded devices are used in place of a multifunctional structural material) [8].

The new functional properties are obtained by increasing the electrical conductivity of the cementitious matrix composites by adding conductive materials such as carbon or metallic admixtures. In any case, structural characteristics should be maintained or improved. Carbon nanotubes (CNT) are one of the additions that can be used to create multifunctional materials. Researching and developing CNT cement matrix composites open a new line to obtain multifunctional materials, possibly more efficient and with a wider range of application fields [9, 10].

Achieving good particle dispersion is the most challenging problem in the fabrication of carbon nanoparticle composites, including CNT cement based composites. Carbon nanotubes have unique mechanical, electrical and thermal properties. However, strong Van Der Waals cohesive forces that occur among the particles results in a high tendency to generate fiber clumps, which are difficult to disperse, and could cause strength losses or even degradation of the original material. The great advances in CNT polymer composites have not been equally achieved in cement composites, probably due to this dispersion problem.

Different methods have been used to disperse CNT or carbon nanofibers (CNF), such as the combined use of surfactants and sonication, the modification of the nanotube surface or functionalizing (incorporating molecular groups to the fiber's surface to improve the affinity with the matrix) [11, 12] and even the direct growth of CNF on unhydrated cement particles [13].

Despite the potential advantages of CNT cement based composites, not many studies have been conducted in order to investigate the mechanical properties of Portland cement mortars fabricated with addition of such nanocomposites [14]. There is a wide variety of results concerning mechanical properties: examples of either high increments [15], or no improvement at all [16] can be found. Undoubtedly, the type of CNT and its dispersion in the cement matrix must play an important role.

On the other hand, although reinforcement corrosion is probably the most important pathology in concrete, no reference to corrosion of steel rebars embedded in CNT cement composites has been found in the available scientific literature. Prior work on cement composites with addition of carbon fibers or other particles showed the importance of admixtures (added to improve fiber's dispersion) on the levels of corrosion measured, which normally were slightly higher due to the enhance composite's conductivity [17, 18]. Consequently, one of the objectives of the present research is the study of the corrosion kinetics of steel reinforcements embedded in cement pastes under aggressive conditions, such as carbonation and chloride attack. To this purpose, in order to characterize its durability, the corrosion of steel reinforced CNT cement pastes with different CNT dosages and subjected to accelerated carbonation and the action of seawater was investigated.

Table 1 Properties of Baytubes® C 70 P Multiwall carbon nanotubes (CNT)

Properties	Value	Unit
C-purity	>95	%
Free amorphous carbon	–	–
Outer mean diameter	~13	nm
Inner mean diameter	~4	nm
Length	>1	μm
Bulk density	45–95	kg/m^3
Elastic modulus	3596	MPa
Tension at break	72.9	MPa
Elongation at break	10.7	%
Izod-impact at 23 °C	103	J/m

2 Experimental

2.1 Materials

Cement mortars were used for mechanical tests (compressive and bending strengths), porosity and density. Cement pastes were used for corrosion rate tests (in order to maximize the influence of the cement matrix on the corrosion process). The materials used in this research were: Portland cement type EN 197-1 CEM I 52.5 R (CEMEX España S.A., Madrid, Spain); Multiwall Carbon Nanotubes (MWCNT, BAYTUBES C 70P), supplied by Bayer Material Science, A.G. (Leverkusen, Germany), whose main properties are listed in Table 1; distilled water; fine aggregate, standard CEN silica sand (in cement mortars); Sikament FF commercial superplasticizer (SIKA, Inc., Valencia, Spain).

The water/cement ratio (w/c) for all mortars and pastes was 0.5, and the cement/sand ratio (c/s) for mortars was 1/3. CNT dosages were 0, 0.05, 0.1, 0.25 and 0.5% by cement mass. The same dosages of CNT and superplasticizer were used for mortars and pastes, and were previously assessed according to Spanish Standard UNE 83258:2005 [19] in order to obtain the same workability for all mortars. Thus the quantities of plasticizer were 0, 0.4, 0.5, 0.9 and 2.2% of the cement mass, for CNT dosages of 0, 0.05, 0.1, 0.25 and 0.5%, respectively.

CNT dispersions for mortars and pastes were done according to a previously checked method in polymer composites [20]. CNT and distilled water were mixed in a rotatory flat-blade mixer and then treated using an ultrasound device model Hielschier UP200S. The resulting mixture was mixed with cement (and sand, in mortars) and superplasticizer in a laboratory planetary mixer for 5 min. Mortars and pastes were fabricated in laboratory conditions: 20 °C temperature and 65% relative humidity (RH). This dispersion method has been successfully used in CNF cement composites [6, 21].

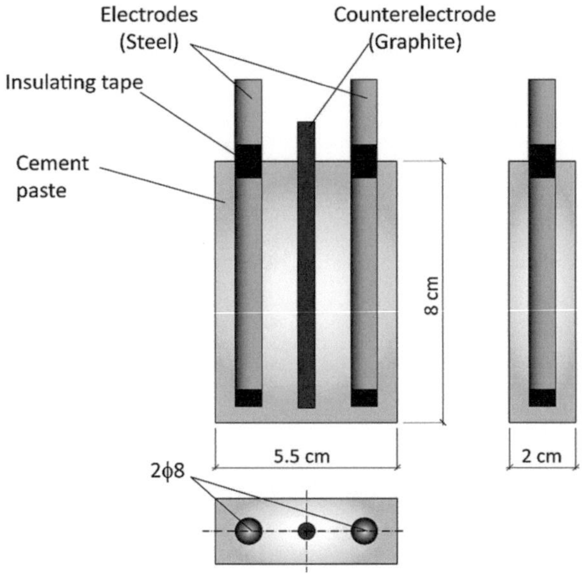

Fig. 1 Dimensions of the specimens for corrosion tests

Prismatic specimens of $4 \times 4 \times 16$ cm^3 were fabricated according to European Standard UNE EN 196-1:2005 [22] for mechanical tests in mortars. They were cured in water until the moment they were tested (at 7 and 28 days curing ages). Mechanical tests were accomplished in laboratory conditions according to UNE EN 196-1:2005. Prismatic specimens of $80 \times 55 \times 20$ mm^3 were prepared for corrosion rate tests. Each one contained two 8 mm diameter cylindrical steel electrodes and a graphite counter electrode in the middle. The thickness of the resulting cement paste cover was 6 mm and the exposed steel area was 16.3 cm^2. Figure 1 shows the specimen arrangement used for corrosion tests, similar to others used in previous works [23, 24].

2.2 Tests Description

Bending and compressive strength tests on prismatic specimens were conducted according to European Standard UNE EN 196-1:2005 [22], with a ME-402/20 press machine (Servosis, S.L., Madrid, Spain). Porosity (P) and apparent density (D_{ap}) were calculated after measuring dry mass (M_d), submerged mass (M_w) and saturated mass (M_s), according to Eqs. (1) and (2). Six samples of each CNT dosage were tested.

$$P(\%) = (M_s - M_d)/(M_s - M_w) \times 100 \quad (1)$$

$$D_{ap} = M_d/(M_s - M_w) \quad (2)$$

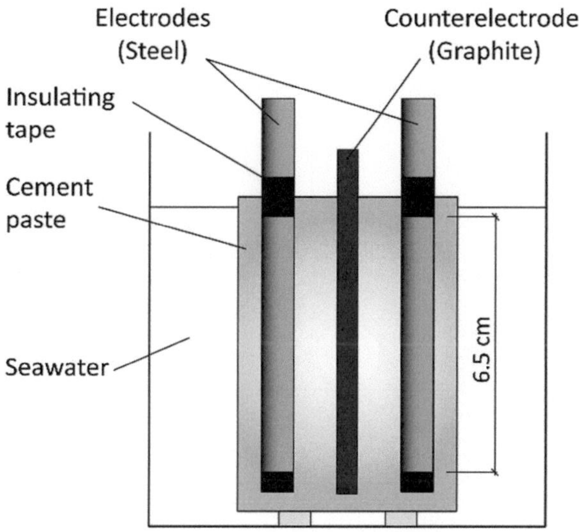

Fig. 2 Specimen partially immersed in seawater for chloride attack tests

Corrosion rate (I_{corr}) and corrosion potential (E_{corr}) were measured in each steel electrode, two electrodes were measured for each specimen. Each process was long enough to obtain steady I_{corr} values.

After curing in ambient-controlled room (100% RH and 20 °C), one specimen of each dosage was immersed in a 0.5 M NaCl solution, simulating seawater. All these tests were done at 20 °C temperature. A part of the specimens was partially immersed in the solution, leaving 1 cm above the water level to avoid direct contact of the electrodes with the solution, as shown in Fig. 2. The other samples were exposed to an accelerated carbonation process in 100% CO_2 atmosphere and 65% ± 5% RH. Polarization resistance technique was used for testing all samples, and instant corrosion rate (I_{corr}) was calculated using Geary and Stern equation [25]:

$$I_{corr} = B/R_p \tag{3}$$

where I_{corr} is the corrosion rate (μA/cm^2); R_p is the polarization resistance (kΩ cm^2) and B (mV) is a constant, assumed equal to 26 mV for the steel-cement system. I_{corr} and E_{corr} were periodically tested. All the potential values were referred to the saturated calomel electrode (SCE). In order to determine the polarization resistance a 362 EG&G potentiostat (Princeton Applied Research) was used. At the end of the experiment, each steel rebar was removed and their gravimetric weight loss determined. The electrochemical weight loss values were estimated by integrating I_{corr} versus time curves and results were compared to the corresponding gravimetric losses (obtained directly by mass differences). The good agreement between both results validates the assumed B value.

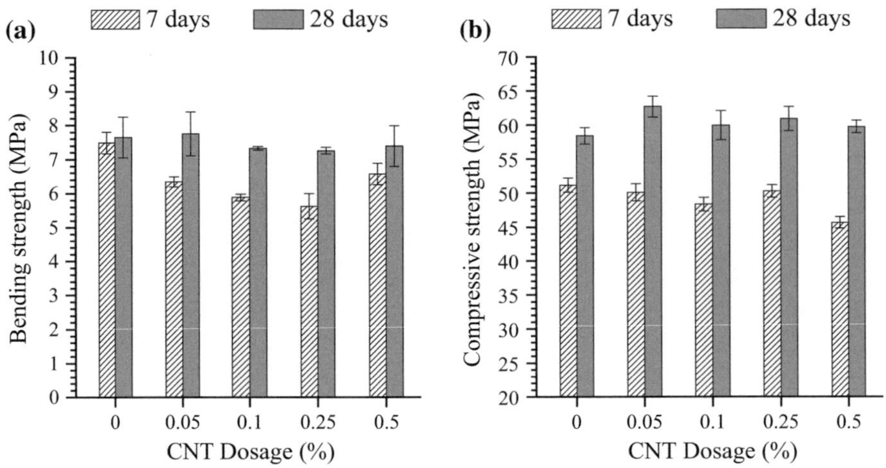

Fig. 3 **a** Bending strength and **b** compressive strength of mortars with different dosages of CNT tested at 7 and 28 days curing time. The indicators represent standard deviations

3 Results and Discussion

3.1 Characterization Tests of CNT Cement Mortars

Figure 3 shows the bending and compressive strengths for each CNT dosage of cement mortars cured in water for 7 and 28 days. Regarding the bending strength, Fig. 3a, all mortars with CNT showed lower strength than the control samples, without CNT, for a 7 days curing time. The increase in CNT dosage did not significantly affect bending strength after a 28-days curing age. Figure 3b shows the influence of the quantity of CNT addition on the compressive strength, for Portland cement mortars tested at curing ages of 7 and 28 days. No significant variations can be observed due to the addition of CNT.

Figure 4 shows the influence of the quantity of CNT addition in the apparent density and porosity for each CNT dosage, in Portland cement mortars at the ages of 7 and 28 days. Mortar density did not significantly change with the content of CNT. However, a slight increase in sample porosity was registered due to CNT addition (between 6.8 and 13.1% with respect to samples without CNT).

As summary of the change in mechanical properties due to different CNT dosage addition, Table 2 shows the increase in mechanical properties of CNT cement mortars with respect to plain cement (without CNT). The highest variation is found for the bending strength at the age of 7 days, as previously mentioned.

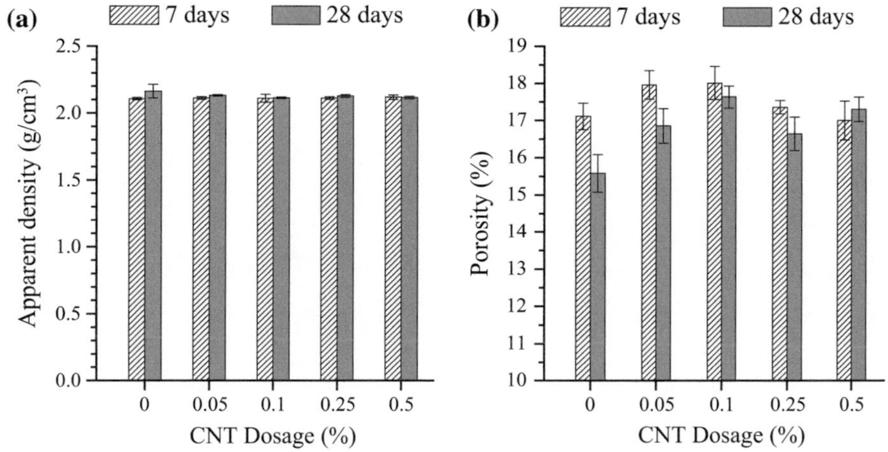

Fig. 4 **a** Apparent density and **b** porosity of mortars with different dosages of CNT tested at 7 and 28 days curing time. The indicators represent standard deviations

Table 2 Increase in mechanical and physical properties (%) after incorporating CNTs

% CNT	Bending strength		Compressive strength		Apparent density		Porosity	
	7 d	28 d	7 d	28 d	7 d	28 d	7 d	28 d
0.05	−15.4	1.3	−2.1	7.3	0.2	−1.5	5.0	8.1
0.10	−21.4	−4.3	−5.5	2.6	0.1	−2.3	5.9	13.1
0.25	−25.0	−5.1	−1.7	4.2	0.2	−1.7	1.4	6.8
0.50	−12.3	−3.4	−10.7	2.3	0.6	−2.2	−0.7	11.0

3.2 Corrosion Tests of Embedded Steel Rebars in CNT Cement Pastes

Reinforced concrete rebars are usually passivated due to the high pH level provided by the cement hydration products. Nevertheless, some circumstances can dissolve the passive film on the steel surface. The main processes leading to the destruction of the passive film are the carbonation of concrete cover and/or the presence of chlorides over the critical concentration threshold. The consequence of transition from passive to active corrosion state is a rapid decrease of the structural element service life.

The aim of this study was the characterization of the corrosion rate observed in embedded rebars in cement pastes with addition of different dosages of CNT. The specimens have been exposed to different aggressive environments. One of these environments has been accelerated carbonation. On the other hand, the chloride ion depassivating action has been studied by partial immersion of the specimens in a solution simulating seawater.

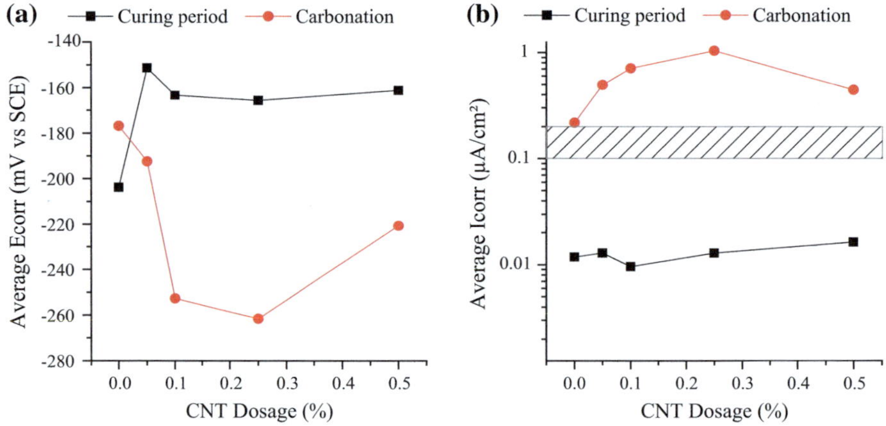

Fig. 5 Average corrosion potential (E_{corr}) and average corrosion rate (I_{corr}) of steel rebars embedded in CNT cement pastes exposed to accelerated carbonation

Figure 5 depicts average values of corrosion potential (E_{corr}) and corrosion rate (I_{corr}) of steel rebars embedded in CNT cement pastes, exposed to the two consecutive exposition conditions: curing period and accelerated carbonation. Cement pastes have different contents of CNT: 0, 0.05, 0.10, 0.25 and 0.50% with respect to cement mass. The hash-marked range in the corrosion rate between 0.1 and 0.2 $\mu A/cm^2$ has been included, showing the threshold between active corrosion (above 0.2 $\mu A/cm^2$) and passivity (below 0.1 $\mu A/cm^2$) [25]. As expected, during the curing period the steel corrosion parameters are consistent with the formation of the passivating layer on the steel surface, and I_{corr} is below the corrosion threshold. A high increase in the corrosion rate was observed as soon as the carbonation process began, and the average I_{corr} is located above the corrosion threshold for all dosages tested. This phenomenon was related to the depassivation of the steel surface due to a pH decrease because of the carbonation of the cement cover. The decrease in the corrosion potential implied the corrosion rate increase. The increase in CNT dosages implied higher final corrosion rates.

Figure 6 shows the corrosion rate (I_{corr}) of steel rebars embedded in CNT cement pastes partially immersed in seawater. At first stage, before the chloride attack, the specimens were stored at 100% RH and 20 °C to obtain an adequate development of the cement matrix. As in the carbonation tests, a correct passivation state was achieved for all specimens for the curing period. After the curing period, the specimens were partially immersed in a 0.5 M NaCl solution simulating seawater chloride concentration. Increases in I_{corr} values were then observed, obtaining a mean value located in the active corrosion area for all specimens, excepting the reference sample (without CNT), that showed the lowest corrosion rate.

Several aspects should be considered to explain the different behavior shown by the steel rebars embedded in cement pastes with different CNT contents. The first one is related to the higher porosity registered for CNT cement mortars than that of

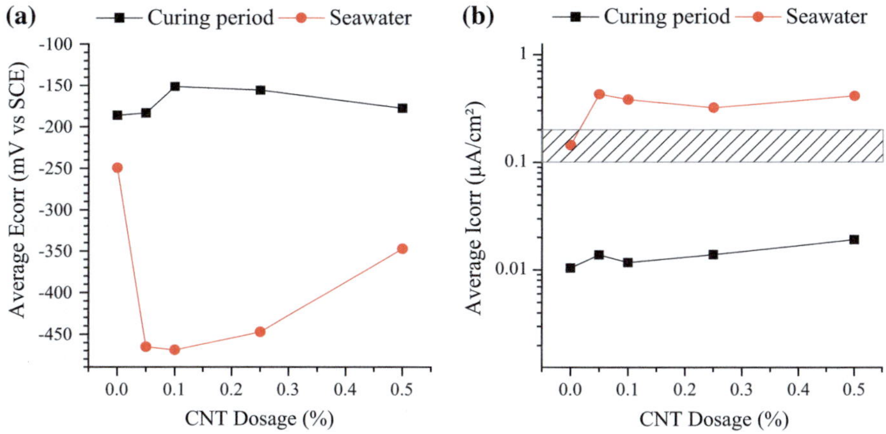

Fig. 6 Average corrosion potential (E_{corr}) and average corrosion rate (I_{corr}) of steel rebars embedded in CNT cement pastes partially immersed in seawater

plain cement mortar (no CNT). Obviously pastes will have a different porosity from mortars, but the former argument can be extrapolated for the corrosion specimens since they have the same cementitious matrix. On the other hand, the increase in CNT addition, which is an electrical conductive material, implies a progressive decrease in the concrete electrical resistivity. This fact contributes to the development of the corrosion cell, which may explain the increase of the I_{corr} values. Finally, the galvanic couple between the steel and the conductive carbon material should be taken into account. The union of two different conductive materials with different nobility implies that the less noble tends to develop higher corrosion rates than the same element without such electrical contact. This argument is consistent because the electrons of the less noble material (steel) would cause cathodic protection on the other one (CNT). For this reason, a higher content of CNT could imply higher levels of the I_{corr} values.

As a consequence of the mentioned factors, it can be expected that the higher CNT dosage would lead to higher conductivity and therefore higher galvanic couple effect from the carbon material to the steel reinforcement. Finally, the complementary action of all three factors implies higher corrosion levels, which is consistent with the obtained results.

4 Conclusions

In the present research, the influence of adding CNT to cement composites was studied in two different aspects. First, some mechanical properties of CNT cement mortars, and second, corrosion rates of steel rebars embedded in CNT cement pastes, were studied. The following conclusions could be drawn.

- The addition of CNT to Portland cement mortars does not significantly affect the bending strength (less than 6%) or the compressive strength (less than 8%), at 28 days curing time.
- The addition of CNT to Portland cement mortars does not significantly affect the apparent density, at 28 days curing time. Only slight increases in the porosity of CNT cement mortars can be detected at the same age (between 6.8 and 13.1% with respect to the control sample without CNT).
- The addition of CNT to the cement matrix could imply the development of higher levels of corrosion in aggressive conditions, such as carbonation and contamination by chloride ions.

Acknowledgements The authors would like to acknowledge the Spanish Ministry of Science and Innovation (Ref: Mat 2009-10866) and Generalitat Valenciana (PROMETEO/2013/035) for their economic support on this research. The authors also thank to Bayer MaterialScience, S.A. for the supply of carbon nanotubes used in this investigation.

References

1. Han B, Sun S, Ding S, Zhang L, Yu X, Ou J (2015) Review of nanocarbon-engineered multifunctional cementitious composites. Compos Part A-Appl S 70:69–81
2. Canon A, Garces P, Climent MA, Carmona J, Zornoza E (2013) Feasibility of electrochemical chloride extraction from structural reinforced concrete using a sprayed conductive graphite powder-cement paste as anode. Corros Sci 77:128–134
3. Gomis J, Galao O, Gomis V, Zornoza E, Garces P (2015) Self-heating and deicing conductive cement. Experimental study and modeling. Constr Building Mater 75:442–449
4. Zornoza E, Catala G, Jimenez F, Andion LG, Garces P (2010) Electromagnetic interference shielding with Portland cement paste containing carbon materials and processed fly ash. Mater Constr 60(300):21–32
5. Chung DDL (2012) Carbon materials for structural self-sensing, electromagnetic shielding and thermal interfacing. Carbon 50(9):3342–3353
6. Baeza FJ, Galao O, Zornoza E, Garces P (2013) Multifunctional cement composites strain and damage sensors applied on reinforced concrete (RC) structural elements. Materials 6(3):841–855
7. Galao O, Baeza FJ, Zornoza E, Garces P (2014) Strain and damage sensing properties on multifunctional cement composites with CNF admixture. Cem Concr Comp 46:90–98
8. Chung DDL, Wen S (2007) Partial replacement of carbon fiber by carbon black in multifunctional cement-matrix composites. Carbon 45(3):505–513
9. Kang IP, Heung YY, Kim JH, Lee JW, Gollapudi R, Subramaniam S, Narasimhadevara S, Hurd D, Kirikera GR, Shanov V, Schulz MJ, Shi DL, Boerio J, Mall S, Ruggles-Wren M (2006) Introduction to carbon nanotube and nanofiber smart materials. Compos Part B-Eng 37(6):382–394
10. Yazdanbakhsh A, Grasley Z, Tyson B, Abu Al-Rub RK (2010) Distribution of carbon nanofibers and nanotubes in cementitious composites. Transp Res Rec (2142):89–95
11. Parveen S, Rana S, Fangueiro R (2013) A review on nanomaterial dispersion, microstructure, and mechanical properties of carbon nanotube and nanofiber reinforced cementitious composites. J Nanomater 2013(710175):19 p

12. Chen SJ, Collins FG, Macleod AJN, Pan Z, Duan WH, Wang CM (2011) Carbon nanotube–cement composites: a retrospect. IES J Part A Civ Struct Eng 4(4):254–265
13. Nasibulina LI, Anoshkin IV, Shandakov SD, Nasibulin AG, Cwirzen A, Mudimela PR, Habermehl-Cwirzen K, Malm JEM, Koltsova TS, Tian Y, Vasilieva ES, Penttala V, Tolochko OV, Karppinen MJ, Kauppinen EI (2010) Direct Synthesis of carbon nanofibers on cement particles. Transp Res Rec 2142:96–101
14. Siddique R, Mehta A (2014) Effect of carbon nanotubes on properties of cement mortars. Constr Build Mater 50:116–129
15. Sobolkina A, Mechtcherine V, Khavrus V, Maier D, Mende M, Ritschel M, Leonhardt A (2012) Dispersion of carbon nanotubes and its influence on the mechanical properties of the cement matrix. Cem Concr Comp 34(10):1104–1113
16. Collins F, Lambert J, Duan WH (2012) The influences of admixtures on the dispersion, workability, and strength of carbon nanotube-OPC paste mixtures. Cem Concr Comp 34(2):201–207
17. Garces P, Fraile J, Vilaplana-Ortego E, Cazorla-Amoros D, Alcocel EG, Andion LG (2005) Effect of carbon fibres on the mechanical properties and corrosion levels of reinforced portland cement mortars. Cem Concr Res 35(2):324–331
18. Garces P, Andion LG, De la Varga I, Catala G, Zornoza E (2007) Corrosion of steel reinforcement in structural concrete with carbon material addition. Corros Sci 49(6):2557–2566
19. "Admixtures for Concretes, Mortars and Grouts. Admixtures for Masonry Mortars. Determination of Consistency"; UNE 83258:2005; Spanish Association for Standardisation and Certification: Madrid, Spain, 2005
20. Bortz DR, Merino C, Martin-Gullon I (2011) Carbon nanofibers enhance the fracture toughness and fatigue performance of a structural epoxy system. Compos Sci Technol 71(1):31–38
21. Galao O, Zornoza E, Baeza FJ, Bernabeu A, Garces P (2012) Effect of carbon nanofiber addition in the mechanical properties and durability of cementitious materials. Mater Constr 62(307):343–357
22. "Methods of Testing Cement–Part 1: Determination of Strength"; UNE-EN 196-1:2005; Spanish Association for Standardisation and Certification: Madrid, Spain, 2005
23. Zornoza E, Garces P, Paya J (2008) Corrosion rate of steel embedded in blended Portland and fluid catalytic cracking catalyst residue (FC3R) cement mortars. Mater Constr 58(292):27–43
24. Garcés P, Andrade C, Saura P, Pérez JM, Zornoza E (2008) Effect of nitrite in corrosion of reinforcing steel in neutral and acid solutions simulating the electrolytic environments of micropores concrete in the propagation period. Corros Sci 50(2):498–509
25. Andrade C, Gonzalez JA (1978) Quantitative measurements of corrosion rate of reinforcing steels embedded in concrete using polarization resistance measurements. Mater Corros 29(8):515–519

Advances in Coal Bottom Ash Use as a New Common Portland Cement Constituent

Cristina Argiz, Esperanza Menéndez and Amparo Moragues

Abstract Cement new constituents can contribute to improve the sustainable of the construction materials and also, in many cases, enhance their durability. Coal bottom ash could be one of such new Portland cement constituents. Currently, they are dumped because are considered as wastes in some countries leading to generate an environmental problem. Therefore, in this paper, coal bottom ash used as a main constituent of Portland cements when it is mixed in an optimized proportion with fly ashes is studied. Mechanical and durable characteristics of standardized mortars made of mixes of coal combustion bottom and fly ashes are compared to mortars made of CEM I 42.5 N (EN 197-1:2011). Ash and cement mixes were designed in order to get similar compositions of CEM II/A-V, CEM II/B-V and CEM IV/A (V) cements according to the European standard EN 197-1:2011. Summing up, it can be said that bottom ash and fly ash mixes perform in the same way than common Portland cements made of fly ash.

1 Introduction

The principal binder in concrete is Portland cement and the production of the clinker necessary to produce Portland cement contributes about 5–7% to greenhouse gas emissions, and therefore, is a major responsible in global warming and climate change. Common cements with additions produced according to EN 197-1:2011 are a way to get better sustainable construction materials. The cement constituents more frequently used in Europe are limestone, siliceous fly ash, natural pozzolans and ground granulated blast-furnace slag [1].

C. Argiz (✉) · E. Menéndez
Instituto de Ciencias de la Construcción Eduardo Torroja (CSIC), C/Serrano Galvache 4, 28033 Madrid, Spain
e-mail: cristina.argiz@hotmail.com

C. Argiz · A. Moragues
E.T.S. Ingenieros de Caminos, Canales y Puertos, Universidad Politécnica de Madrid, Ciudad Universitaria, C/Profesor Aranguren s/n, 28040 Madrid, Spain

C. Andrade et al. (eds.), *Service Life and Durability of Reinforced Concrete Structures*, RILEM Bookseries 17, https://doi.org/10.1007/978-3-319-90236-4_4

Because this way addresses all three sustainability issues, its use will enable the cement and concrete industry to become more sustainable, taking into account the definition of sustainable development given in the Brundtland Report, in 1987, as the ability of humanity "to ensure that it meets the needs of the present without compromising the ability of future generations to meet their own needs" [2].

Fly ash is the main by-product of the coal-fired power plants. It is a well-known pozzolanic binder that may be used either as a constituent of common Portland cements or as a supplementary cementitious material (SCM) in concrete. Usually, the amount of fly ash between 15 and 20% by mass of the total cementitious material has a beneficial effect on the workability, improve the durability to sulphate attack, alkali-silica expansion, and so on. Such beneficial effect is a consequence of the pozzolanic reaction where the silicon dioxide, SiO_2, react with the portlandite, $Ca(OH)_2$, which is produced from the calcium silicates (C_3S and C_2S) hydration. Ground bottom ash presents also a pozzolanic performance [3]; however, it is normally used as aggregate [4] without grinding.

Concrete made of fly ash when properly cured is able to provide excellent durability. The mechanisms responsible for this phenomenon may be explained as follow. The area between the aggregate and cement paste called interfacial transition zone has a high water/cement ratio and, therefore, it promotes to have more available space that allows the formation of a highly porous hydration product and large hexagonal crystals of calcium hydroxide and needles of ettringite. Then, microcracks are easily formed through this product because it is weaker than the bulk cement paste. It has been suggested that microcracks in the interfacial transition zone play an important part in determining not only the mechanical properties but also the permeability and durability of concrete. Finally, with the progress of the fly ash pozzolanic reaction, large pores and large crystalline products in the transition zone are reduced [5]. This fact would contribute to enhance the concrete durability and compressive strength at latter ages in well-cured cementitious products [6, 7].

Bottom ash and fly ash are formed together in the boiler of coal fired power plants. Thus, similar chemical composition in both ashes has been reported [3, 8–10]. However, only few studies on bottom ash employed as cement constituent have published and, nowadays, coal bottom ash is normally used as aggregate in road bases [4].

This paper compares the compressive strength and porosity of standard mortars made of coal bottom ash or coal fly ash. Also, this paper proposes that ground bottom ash could be used as a new common Portland cement constituent in the near future.

As the population increases and concrete is used to meet various needs of these people, we must try to reduce the impact of the production and use of concrete.

2 Experimental

2.1 Raw Materials

A CEM I 42.5 N cement according to the European standard EN 197-1:2011 [1] was employed to prepare laboratory blended cements made of fly ash and bottom ash. A German standard sand (NORMSAND) and distilled water were used to make testing mortars.

Fly ash and bottom ash were provided by ENDESA-Carboneras. They came from South African hard coal (90%) and Colombian hard coal (10%). Bottom ash was ground to a fineness equivalent to that of the Portland cement (5–8% residue on 45 μm sieve). Fly ash was used as it was provided.

2.2 Chemical Composition and Physical Characteristics

The chemical composition of the fly ash, bottom ash and cement is shown in Table 1 and Portland cement CEM I 42.5 N (EN 197-1:2011). Chemical analyses of SiO_2, Al_2O_3, Fe_2O_3, CaO, MgO, SO_3, TiO_2, P_2O_5 were performed by XRF with a Bruker S8 Tigger 4 kW model. Loss on ignition (LOI) and insoluble residue (IR) were determined according to EN 196-2:2005. Expansion, initial and final setting times were determined according to EN 196-3:2005+A1:2008. Finally, density and specific surface (Blaine) were determined according to EN 196-6:2010.

Table 1 Chemical compositions of raw materials: bottom ash (BA), fly ash (FA) and cement (as percentage)

Chemical contents (%)	Cement	BA	FA
I.R.	2.15	81.24	76.00
L.O.I.	3.93	1.85	3.63
SO_3	3.14	0.15	0.37
Al_2O_3	3.85	25.55	26.66
Fe_2O_3	3.43	5.86	4.72
CaO	57.16	7.07	5.55
MgO	1.54	1.28	1.33
SiO_2	19.04	48.12	46.84
TiO_2	0.17	1.50	1.50
P_2O_5	0.07	0.96	1.03

*Insoluble residue determined by the Na_2CO_3 method (EN 196-2:2005)

Table 2 Bottom ash (BA) with fly ash (FA) and cement (CEM I 42.5 N) mixes

Cement mix	Raw material	Fly ash + bottom ash mix codification					
		α (%)	β (%)	γ (%)	δ (%)	λ (%)	Ω (%)
CEM I	Fly ash	0					
	Bottom ash	0					
	Cement	100					
CEM II/A-V	Fly ash		10	9	8	5	0
	Bottom ash		0	1	2	5	10
	Cement		90	90	90	90	90
CEM II/B-V	Fly ash		25	22.5	20	12.5	0
	Bottom ash		0	2.5	5	12.5	25
	Cement		75	75	75	75	75
CEM IV/A (V)	Fly ash		35	31.5	28	17.5	0
	Bottom ash		0	3.5	7	17.5	35
	Cement		65	65	65	65	65

2.3 Blended Cement Design

Fly ash and bottom ash mortars mixes were coded as β, γ, δ, λ and Ω (Table 2), and reference cement was coded as α (CEM I 42.5 N). Cement was partially replaced by fly ash, bottom ash or mixes in order to prepare laboratory cements with similar compositions to CEM II/A-V, CEM II/B-V and CEM IV/A (V) standardized cements [1]. Their properties were compared to those of CEM I 42.5 N. Table 3 summarizes the composition of these cement types elaborated with fly ash (β), bottom ash (Ω) or mixes using the reference cement (α) as raw material.

2.4 Experimental Testing

Compressive strength. Compressive strength tests were performed on mortars at 1, 3, 7, 14, 28 and 90 days according to the European standard EN 196-1:2005. An IBERTEST-AUTOTEST 200-10-5 wc model was used to perform the test.

Mercury intrusion porosimetry (*MIP*). Pore size distribution was measured by means of the MIP technique with the help of a Micromeritics AUTOPORE IV 9550. Washburn equation shown in Eq. (1) was used to determine the pore size.

Table 3 Compliance with specifications: Fly ash and bottom ash from Carboneras

Performance characteristics		Carboneras		Standard and regulatory requirements				
		Bottom ash	Fly ash	Fly ash in common cements EN 197-1:2011	Siliceous fly ash (V) in common cements SR UNE 80303-1:2013	Siliceous fly ash (V) in common cements MR UNE 80303-2:2011	Fly ash in concrete	
							EN 450-1: 2013	EHE-08
Reactive silica (SiO_2) (%)		30.38	33.38	≥25	≥25		≥25	–
Loss on ignition (%)		1.85	3.63	0–5.0 2.0–7.0 4.0–9.0	0–5.0 2.0–7.0 4.0–9.0		A ≤ 5.0 B ≤ 7.0 C ≤ 9.0	≤5.0 Category A
Sulfuric anhidride (SO_3) (%)		0.15	0.37				≤3.0	≤3.0
Total calcium oxide (%)		7.07	5.55				≤10.0	
$SiO_2 + Al_2O_3 + Fe_2O_3$ (%)		79.53	78.22				≥70	
MgO (%)		1.28	1.33				≤4.0	
Total phosphate P_2O_5 (%)		0.96	1.03				≤5.0%	
Strength Activity Index (%)	28 days	85%	87%		≥75		≥75	≥75
	90 days	98%	109%				≥85	≥85

$$P = -\frac{4\gamma \cos\theta}{D} \tag{1}$$

where:

γ Mercury superficial tension equal to 485 N/m (dinas/cm).
θ Mercury-pore contact angle (°), usually ranged between 112° and 142°.
D Diameter of the pore filled of mercury (μm).
P Mercury applied pressure (MPa).

Then, pore size distribution can be calculated using Eq. (2) in function of the applied pressure.

$$D = -\left(\frac{1}{P}\right)4\gamma\cos\theta \tag{2}$$

Durability testing. Bottom ash durability performance was considered with regard to sulphate attack, alkali-silica reaction, carbonation resistance and frost-thaw resistance. The procedure was defined elsewhere [11].

3 Results and Discussion

Bottom ash and fly ash chemical specified parameters are shown in Table 3. SiO_2 + Al_2O_3 + Fe_2O_3 is a parameter specified for fly ash. For the bottom ash this value is 79.53% and for the fly ash 78.22%. Both are higher than ≥70.0% and closer each other, being a Type F fly ash (ASTM C 618-12a). Loss on ignition (LOI) for bottom ash is 1.85% and for fly ash is 3.63%. Then, both ashes fit the requirements of EN 450-1:2012.

3.1 Mechanical Strength

Compressive strength is the characteristic of used in structural design. The effect of bottom ash compressive strength is compared to the fly ash one. In Figs. 1, 2, and 3, compressive strength at 1, 3, 7, 14, 28 and 90 days of tested mortars made of fly ash, bottom ash and mixes is shown. As expected, the highest compressive strength is at 90 days. In particular, ash mortars presented a lower strength at early ages due to the slower rate of the pozzolanic reaction.

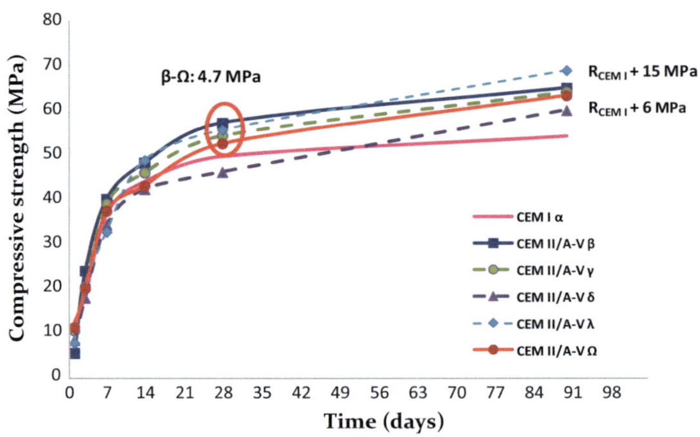

Fig. 1 CEM II/A-V compressive strength (MPa)

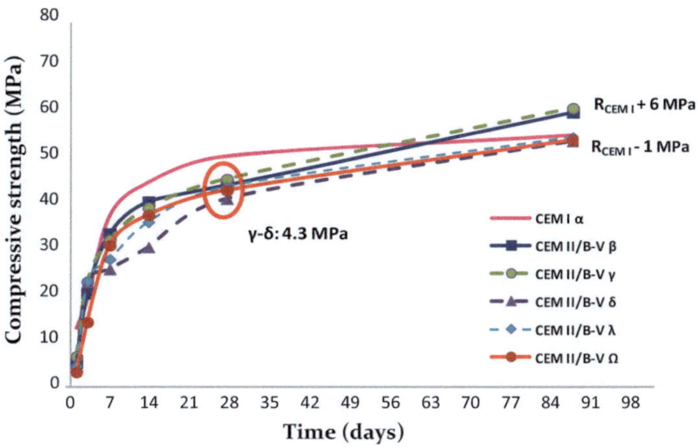

Fig. 2 CEM II/B-V compressive strength (MPa)

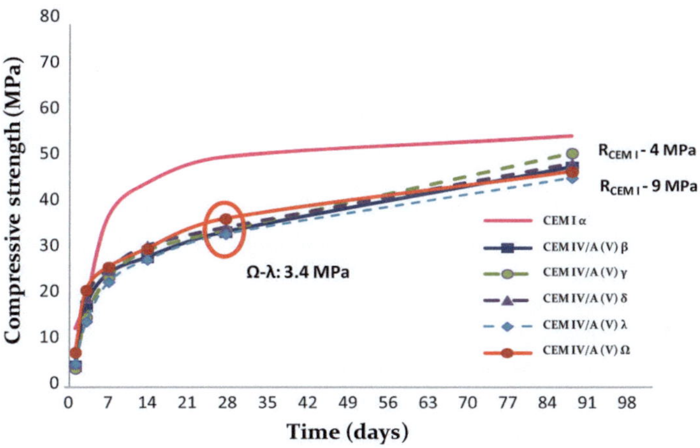

Fig. 3 CEM IV/A (V) compressive strength (MPa)

In Fig. 1 is observed compressive strength evolution of the mortars elaborated with CEM II/A-V cements in comparison to the reference cement. The continuous line without symbol is the compressive strength of the reference mortar. Also, continuous lines with square and circle symbols represent mortars with 100% of fly ash and 100% of bottom ash, respectively. On the other hand, discontinuous lines represent mortars with ash mixes. After 14 days, some mortars have exceeded slightly the compressive strength of the reference mortar. But also, it is observed that all the mortars have a compressive strength higher than that of the reference mortar, with the exception of the mortar with 20% of bottom ash. In this case, it is believed that an anomalous performance has been recorded.

The compressive strength results at 28 days do not show significant differences between the fly ash mortars and bottom ash mortars, or mixes of them. The highest compressive strength at 28-days was above 50 N/mm^2 in CEM I 42.5 N and CEM II/A-V. The highest compressive strength difference of mortars with 100% of fly ash and 100% of bottom ash is 4.7 MPa. This low difference allows saying that there are no significant differences between the proportions of ashes considered in the study. Then, type CEM II/A-V cements can be classified as 42.5 N strength class (2 days ≥ 10 MPa and 28 days ≥ 42.5 MPa).

Finally, all the mortars increase their compressive strength along the time up to 90 days, where all the ash mortars have higher compressive strength than the reference mortar. After 90 days, mortars with 100% of fly ash and with 100% of bottom ash have similar compressive strengths. The compressive strength difference between the ash mortars and reference one is 6 MPa in the case of the mortar with lowest compressive strength and 15 MPa in the case of the mortar with highest compressive strength.

Figure 2 shows the evolution of the mortar compressive strength made of CEM II/B-V cement. In this case, at 14 days, all the CEM II/B-V strength is still lower than in the reference mortar. Difference in compressive strength between fly ash mortars and bottom ash mortars, or mixes of them is 4.3 MPa, lower than the difference of CEM II/A-V mortars and reference mortar.

CEM II/B-V cements can be classified as 32.5 N strength class (7 days ≥ 16 MPa, 28 days ≥ 32.5).

After 90 days, mortar compressive strength with 100% fly ash or 90% bottom ash exceed the reference mortar compressive strength in 6 MPa. Bottom ash mortar with 100% bottom ash (Ω) and the rest of the mortar mixes presented compressive strengths slightly lower than the reference mortar.

Finally, Fig. 3 shows the evolution of the CEM IV/A (V) mortars compressive strength. In these mortars, the compressive strength is very low at 7 days. After 28 days, differences in compressive strength with regard to the reference mortar are larger than in the case of the CEM II/B-V mortars. Nevertheless, differences between fly ash and bottom ash mortars or other mixes are very small. This low compressive strength decreases along the time and after 90 days, the highest difference in compressive strength with the reference mortar is 9 MPa.

After 90 days, there are no significant differences between fly ash and bottom ash mortars or other mixes are very small. CEM IV/A (V) cements can be classified as 32.5R strength class (2 days ≥ 10 MPa, 28 days ≥ 32.5).

From the information presented till now, it is possible to conclude that cements with higher ash contents than 25% CEM II/B-V and CEM IV/A (V) can be more pozzolanics than cements with 10% of ash CEM II/A-V. However, this pozzolanic performance is not reflected providing higher compressive strength.

3.2 Mercury Intrusion Porosimetry (MIP)

Figures 4, 5, and 6 show mercury intrusion porosimetry (MIP) results for all the studied mortars. The evolution of the critical pore diameter, which is the maximum peak of the pore size differential distribution curve it is show in Fig. 4.

It is observed that after 7 days, all the blended mortars present larger critical pore diameters than reference cement. These differences become higher when the amount of addition increases. After 28 days, when the pozzolanic reaction has began in all the mortars, an evolution towards smaller critical pore diameters is observed, in particular, in mortars with 25 and 35% of ash. It is observed that after 90 days of hydration, the pore diameter for all the ash percentages and for the mixes is similar to that of the reference mortar.

The pore diameter distribution of different mortar mixes made of cement CEM II/A-V obtained at the ages of 7 and 28 days, respectively is shown in Figs. 4 and 6. The pore diameter of reference mortar is about 1 μm. After 7 days, the pore diameters of blended mortars have diameters higher to those of the reference mortar. A slight dispersion is observed between the curves. After 28 days, blended mortars decrease his pore diameters becoming closer to the pore size of the reference mortar.

According to the experimental results, the mortar porosity at early ages increase with the content of aches. As hydration time progresses it is produced refinement of pores which is more evident in blended mortars with 25 and 35% of ash. This pore size refinement towards smaller sizes, does not have an effect in compressive strength increase, however, it can provide durability benefits [12].

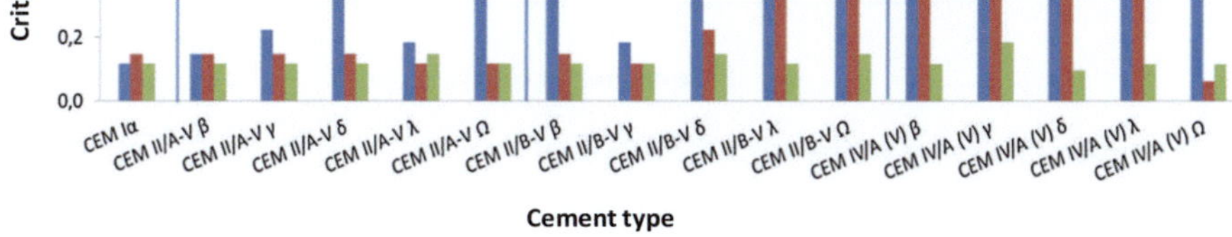

Fig. 4 Critical pore diameter determined by (MIP) from 7 to 90 days

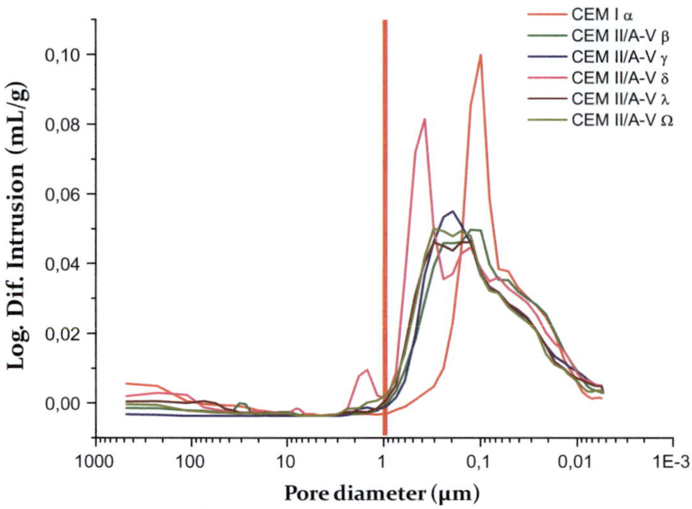

Fig. 5 Mercury intrusion porosimetry (MIP) at 7 days

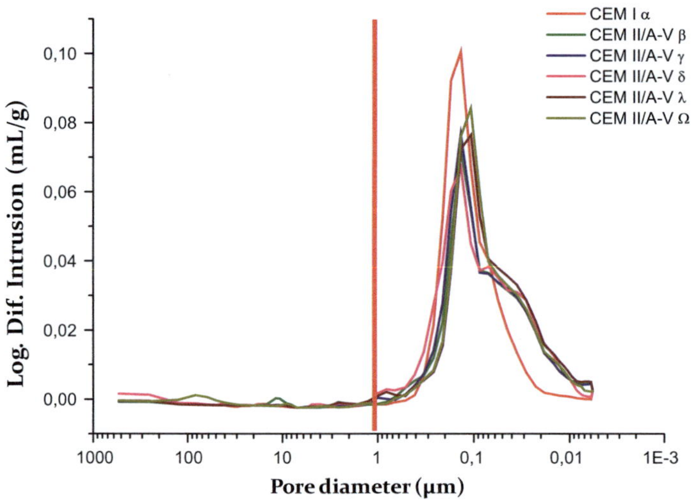

Fig. 6 Mercury intrusion porosimetry (MIP) at 28 days

3.3 Durability

Bottom ash durability performance was considered with regard to sulphate attack, alkali-silica reaction, carbonation resistance and frost-thaw resistance [11].

In general, bottom ash addition improves the mortar resistance to the sulphate attack and alkali-silica reaction. Mortars with higher ash contents (CEM II/B-V and

CEM IV/A (V)) have been recognized as the most resistant, regardless the ash mix used.

On the contrary, bottom ash mortars showed a lower carbonation and frost-thaw resistance. However, small percentages of bottom ash (10%) did not show a significant effect.

4 Conclusion

Coal bottom ash mortars showed compressive strength, carbonation resistance and pozzolanic activity similar to that of fly ash mortars. Therefore, coal bottom ash alone or mixed with fly ash is suitable to be used in cement production because. Summing up, it can be concluded that there are not differences between fly ash and bottom ash either with relation to the mechanical or durable performance. Finally, the data presented here support the incorporation of bottom ash as a main constituent in the Portland cement standards.

Acknowledgements The financial support of the ***Instituto Español del Cemento y sus Aplicaciones (IECA)***, under the Research Contract 'Durable physico-chemical characterization of cement-based materials with industrial by-products'—AU 23-10 Contract, is gratefully acknowledged.

References

1. Sanjuán MA, Argiz C (2012) The new European standard on common cements specifications EN 197-1:2011. Mater Construcc 62:425–430
2. The World Commission on Environment and Development, Our Common Future, p 8
3. Argiz C, Menéndez E, Sanjuán MA (2013) Effect of mixes made of coal bottom ash and fly ash on the mechanical strength and porosity of Portland cement. Mater Construcc 309:49–64
4. Churcill VE, Amirkhanian SN (1999) Coal ash utilization in asphalt concrete mixtures. J Mater Civil Eng 11:295–297
5. Taylor HFW (1997) Cement chemistry, 2nd edn. Thomas Telford Publishing, London
6. Sánchez de Rojas MI, Frías M (1996) The pozzolanic activity of different materials, its influence on the hydration heat in mortars. Cement Concr Res 26:203–213
7. Frías M, Sánchez de Rojas MI (2003) The effect of high curing temperature on the reaction kinetics in MK/lime and MK/cement matrices at 60°C. Cement Concr Res 33:643–649
8. Cheriaf M, Rocha JC, Pera J (1999) Pozzolanic properties of pulverized coal combustion bottom ash. Cement Concr Res 29:1387–1391
9. Argiz C, Menéndez E, Moragues A, Sanjuán MA (2014) Recent advances in coal bottom ash use as a new common Portland cement constituent. Struct Eng Int 4(2014):503–508
10. Menéndez E, Álvaro M, Argiz C, Parra JL, Moragues A (2013) Boletín de la Sociedad Española de Cerámica y Vidrio 52(6):296–304
11. Argiz C (2014) Study on coal-fired power plants bottom ash use as a Portland cement constituent. Comparative analysis regard to fly ash. PhD. Thesis
12. Argiz C, Menéndez E, Moragues A, Sanjuán MA (2015) Fly ash characteristics of Spanish coal-fired power plants. Afinidad 72:572

Part II
Techniques of Characterization of Corrosion and Degradation

Numerical Simulations for the Detection of Leakages in Bridge Deck Membranes Through Resistivity Measurements

Carla Driessen and Michael Raupach

Abstract Through measurement of the electrolytic resistance between two carbon meshes embedded in a textile reinforced concrete interlayer on bridges, a large scale monitoring of humidity is implemented. The electrolyte resistance is measured using alternating current. Decreasing values in the measured electrolyte resistance indicate water ingress through leakages in a surface-applied membrane. The monitoring will have a positive effect on the durability of a bridge construction since defects in the membrane can be detected in an early stage which allows cost-effective and timely measures to be taken to prevent corrosion initiation of the reinforcing steel. Through numerical simulations the theoretical size of a detectable leakage area is calculated which depends on the geometry of the leakages, the electrolyte resistance of the mortar in wet and dry condition, the position of the leakage, and other boundary conditions.

1 Detection of Leakages Through Resistivity Measurements

Leakages in membranes on concrete bridge decks normally threaten the durability of the bridge structure as water and dissolved chloride ions can easily permeate into the concrete cover which may eventually lead to corrosion of the steel reinforcement. In practice leakages often remain largely unnoticed until corrosion damages can be observed on the external surfaces of a bridge. In this stage it is already very expensive to repair these damages of the structure. For a more proactive maintenance of bridges the idea of a full surface humidity-monitoring was developed, which is intended to transmit the relevant data in real time. Until now, only sensors which measure selectively at the point of installation are available [1, 2]. For a full surface humidity-monitoring the resistance between two carbon meshes having an in-between spacing of 15 mm is measured. The measured resistance can then be

C. Driessen (✉) · M. Raupach
Institute of Building Materials Research, RWTH
Aachen University, Schinkelstrasse 3, 52062 Aachen, Germany
e-mail: Driessen@ibac.rwth-aachen.de

C. Andrade et al. (eds.), *Service Life and Durability of Reinforced Concrete Structures*, RILEM Bookseries 17, https://doi.org/10.1007/978-3-319-90236-4_5

converted into a resistivity which allows a quantification of the moisture content as to assess the occurrence of a leakage. The resistivity of concrete may vary over a wide range with resistivity values of about 10^6 Ωm for dry concrete whereas water-saturated concrete is characterized by an electrolyte resistivity of approximately 100 Ωm [3].

1.1 Relationship Between Moisture Content and Resistivity

Knowing the exact relationship between moisture content and resistivity values of a specific mortar and by detecting a change in the electrolyte resistance allows to estimate the amount of water which has permeated into the structure. The relationship between moisture content and mortar resistivity can be investigated in laboratory tests under well-defined conditions.

For these laboratory investigations mortar specimens are stored under water until full water saturation is achieved. In this condition the electrical resistance and the weight of the specimens are measured. The resistance is measured with a two electrode set-up. Thereupon the specimens are dried at 70 °C until low moisture contents are obtained. When such a condition has been achieved, a wide variety of moisture contents are being adjusted through the addition of different amounts of water. Following this treatment the specimens are packed vapour resistant to reach a homogenous distribution of the water. After a few weeks the specimens are unpacked and weighed. In addition, the electrolyte resistance is being measured. Eventually the specimens are dried at 105 °C until a constant weight is achieved and this dry weight is measured. Based on these data the relationship between moisture content and resistivity values can be calculated [4]. This requires the conversion of an AC resistance into a resistivity which is achieved using a cell constant according to Eq. (1) [5].

$$\rho = \mathrm{R_{el}} \cdot \frac{\mathrm{A}}{\mathrm{l}} \tag{1}$$

with:

ρ mortar resistivity in Ωm
R_{el} measured resistance in Ω
A cross sectional area of the mortar specimen in m^2
l distance between the electrodes in m

Having determined the resistivity values with the corresponding moisture contents, a regression analysis is carried out using the following regression function (2) [6]:

$$u = \frac{A}{\ln(B + \rho + C) - D} + E \tag{2}$$

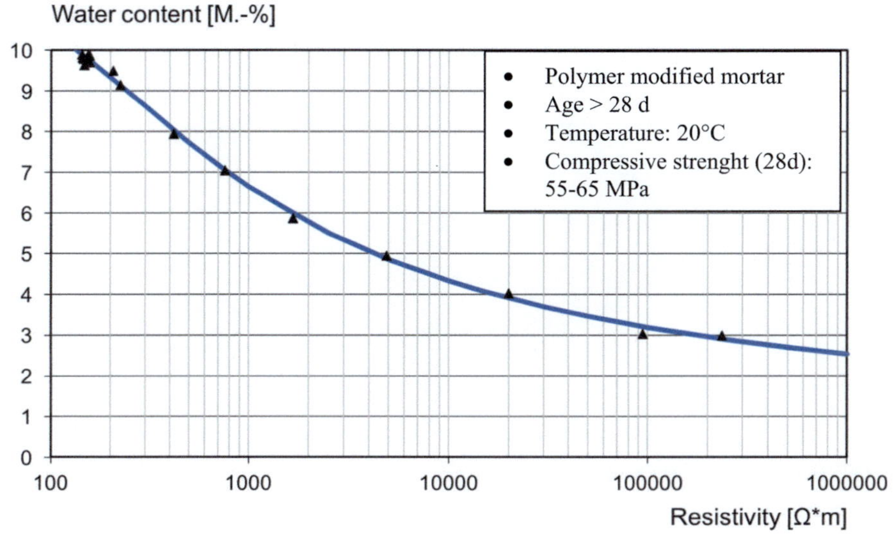

Fig. 1 Calculated relationship between water content and resistivity

with:

u moisture content in M.-%
ρ mortar resistivity in Ωm
A–E regression parameters

This regression curve is based on an empirical formula derived by Raupach [7]. Figure 1 shows a graph of the measured values and the calculated regression curve. This regression curve serves as the main input value for the numerical simulations.

For the numerical simulations values ranging from x to 1000 Ωm represent the wet mortar in the leakage zone and values above 1000 Ωm are representative for the mortar in a dry state under an intact membrane. Of course there is not a clear/distinct threshold level which separates between wet and dry condition of a mortar but a gradual transition depending on the conditions for desiccation such as temperature, humidity and the presence of a membrane. However, these influencing parameters were not part of the numerical simulations but will be examined in laboratory tests in a later stage of the project. Preliminary tests show that after a few months of desiccation the specimens reach water contents between 6 and 7 M.-%, which is demonstrated by resistivity values of about 1000 Ωm.

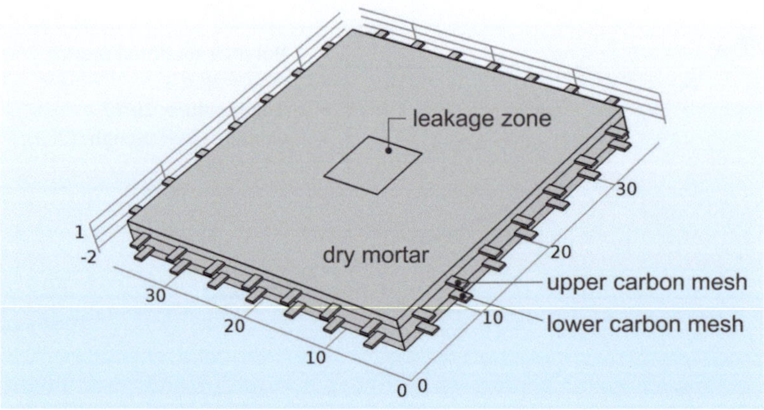

Fig. 2 Basic model with a leakage zone in the centre

2 Numerical Model and Parameter Sweep

The interlayer arrangement which will be applied on bridge decks [4] was transferred to specimens with a cross section of 30 by 30 cm^2 for laboratory testing. These specimens serve as the model for the numerical simulations. This downscaled model enables shorter calculation times. Under real conditions the measurement areas on bridge decks will have a size of about 5–10 m^2. In the following the results will be given in relative sizes as to transfer the results to all possible geometries. The interlayer arrangement is built up of a 10 mm mortar bottom layer, serving as the contact zone to the matured concrete. On this bottom layer two carbon meshes with a in-between spacing of 15 mm are placed. Both carbon meshes are embedded in a cement-based mortar. The top layer consists again of 10 mm mortar. Both carbon meshes have a mesh size of 38 mm. The basic model is shown in Fig. 2 with a leakage zone in the centre.

Between the upper and bottom carbon mesh a fixed AC voltage is applied with the lower carbon mesh connected to ground. In the following the system of the two carbon meshes is called "anode" to distinguish it from the reinforcement steel in the bridge deck which acts as the cathode, however this is not treated in the present paper. The voltage applied can be chosen arbitrarily because it will be offset by the resulting current because it is assumed that no polarisation of the carbon meshes occurs. This assumption is justified because the resistivity will be measured with alternating current. The resistivities for the mortar in wet and dry condition are the input values for the numerical simulations, resulting into respective values for the conductance as an output. In order to determine the resistivity first the geometry factor for the system with the mesh anode was calculated being dependent on both, respectively, the specimen size and the roving quantity. The resulting regression curve for the geometry factor is depicted in Fig. 3. This conversion from a resistance

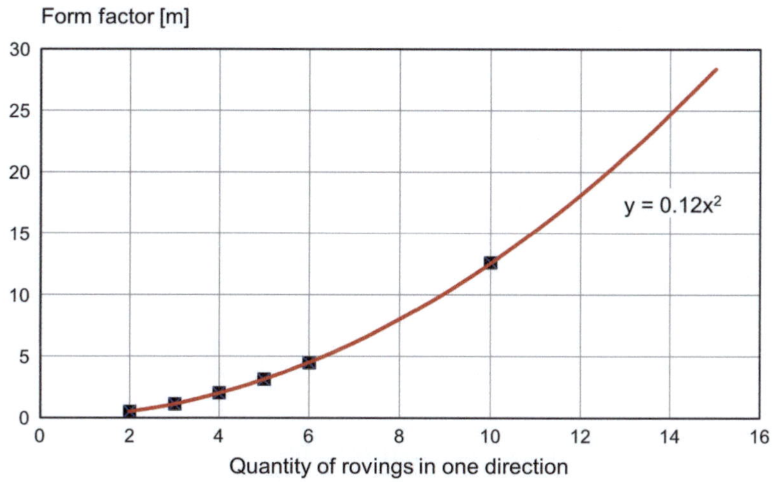

Fig. 3 Form factor for the numerical model depending on the model size

into a resistivity is essential for comparison of the numerical results because only the resistivity is corrected for geometry effects.

The majority of the numerical simulations was carried out with the arrangement and geometry of the model shown in Fig. 2, which corresponds to a roving quantity of seven in each direction and a geometry factor of 6.15 m. Moreover, for comparison, numerical simulations with a model size of 10 m^2 were calculated. The results of both models given in relative sizes matched with a negligible difference of 0.17%. Thus the assumption that the results can be transferred to all possible measuring fields seems to be correct.

2.1 *Investigated Parameters*

The main issue to be resolved through numerical simulations is which size of a local defect in a membrane can be detected at which resistivity values of the wet mortar located in the leakage zone and at which resistivity values of the dry mortar under an intact membrane. Moreover the shape and the depth as well as the position of the leakage in the measuring field could influence the measured drop in the resistivity values. Another question is whether there is a difference between the situation in which just a single leakage at a time occurs and the situation that numerous smaller leakages occur nearly at the same time which are distributed over the measuring field. In addition it was examined if the simplification of the mesh anode to a flat full surface anode is allowed with respect to the resulting error. In Table 1 provides an overview of the varied parameters.

Table 1 Varied parameters

	Dimensions	Range
Size of the leakage zone	% of the total field	1–100
Resistivity of the leakage zone	Ωm	150; 250; 500; 700
Resistivity of the dry mortar	Ωm	1000; 5000; 7000; 100,000
Depth of the leakage	cm	1–3
Position of the leakage	–	Center; edge
Number of leakages	–	1; 4; 9
Shape of the leakage	–	Cubic; cylindrical
Shape of the anode	–	Mesh; flat

3 Results of the Numerical Simulations

First, the spatial distribution of the electrical current expressed as a current density and the flow direction of the applied current were plotted. Due to the relatively low resistivity of the leakage zone it is logical that the current flows in the direction of the leakage zone, as exemplified in Fig. 4.

Accordingly the current densities in areas nearer to the leakage zone are higher. This can clearly be observed in Fig. 5 on the left side which shows a view from above. On the right side one can see a sectional view between the two carbon meshes. Only in the leakage zone increased current densities appear, so the majority of the applied current will flow through the leakage.

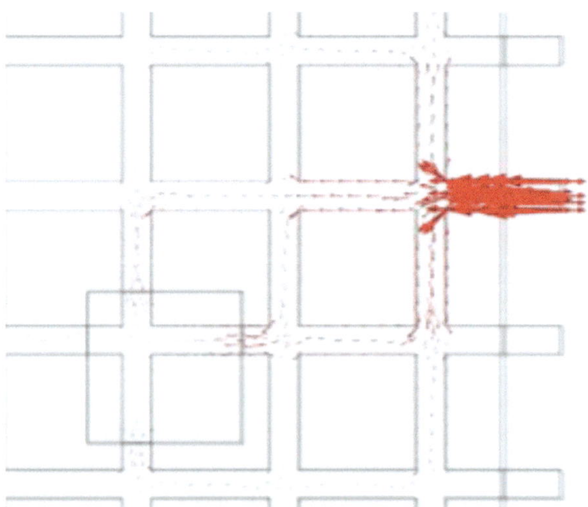

Fig. 4 Flow of the applied electrical current

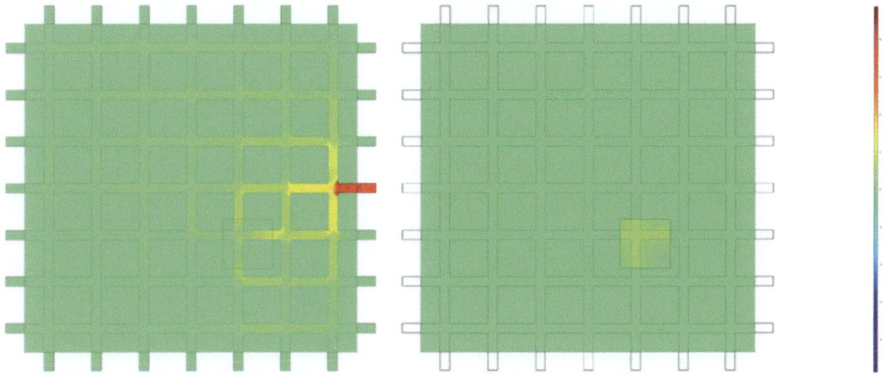

Fig. 5 Current densities shown from above (left) and between the carbon meshes (right)

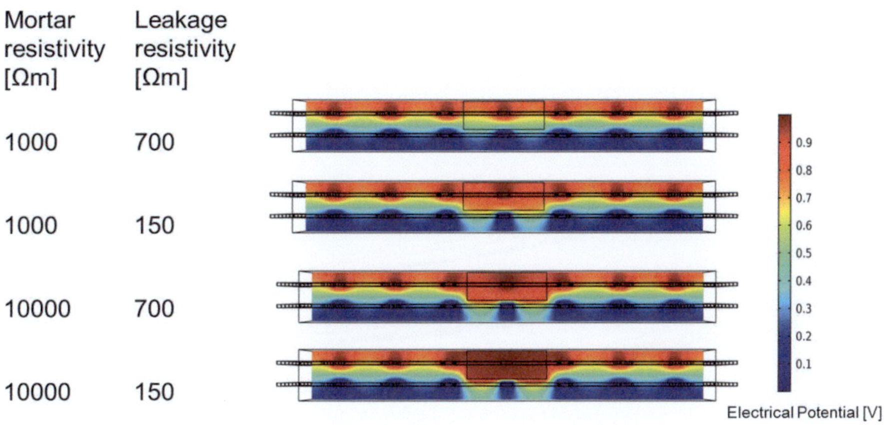

Fig. 6 Distribution of the potential for different mortar resistivity combinations

3.1 Influence of Different Resistivities and Sizes

In the first simulations the resistivities for wet and dry conditions of the mortar as well as the size of the leakage were varied. Figure 6 shows four different combinations of resistivity values starting with the smallest difference between resistivity values in wet and dry condition up to the greatest difference shown at the bottom.

As an example Fig. 7 shows the calculated drop in mortar resistivity for various leakage resistivities based on a dry mortar resistivity of 1000 Ωm under an intact membrane. This is the resistivity which can theoretically be derived with the measuring technique according to the numerical simulations.

The larger the leakage the easier it will be detected by the decrease in resistance. Secondly, the lower the resistivity of the leakage zone the easier the leakage can be found. From Fig. 4 is can be deduced that leakages with the size of a few percentages

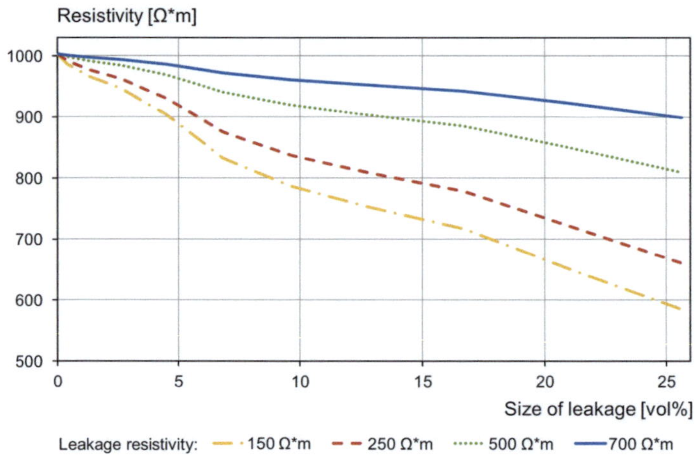

Fig. 7 Drop in resistivity values caused by different mortar resistivities in the leakage zone

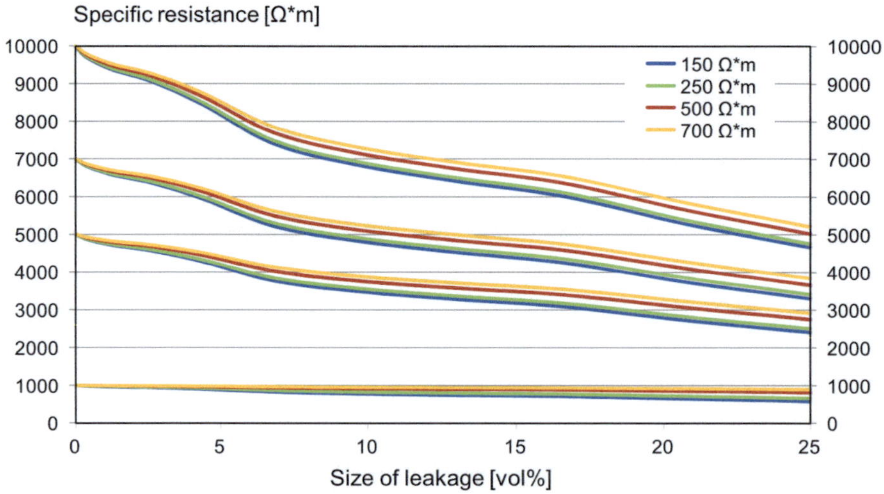

Fig. 8 Drops in resistivity values

of the total field size are detectable. The depicted case is the one with the toughest conditions for the detection of leakages being investigated. All other cases show higher resistivities of the dry mortar and therefore larger drops in resistivity caused by the leakages. For better comparison the resistivity is shown in Ωm. The resistance in Ω depends on the field size. The smaller the measuring field the larger are the measured drops and the easier leakages are detected. All curves show some bends caused by the mesh structure of the anode.

Figure 8 shows the decrease of the resistivity for different resistivity combinations of wet and dry mortar.

Figure 8 is not clear: 150, 250, 500 and 700 refer to leakage resistivities. To what does the specific resistance refer to? Of the combination of the leakage zone and the unaffected dry zone; does the Y-axis indicate an apparent resistivity?

It becomes obvious that the resistivity of the dry measuring field has a higher impact on the detectability of leakages than the resistivity of the leakage zone. The differences between the different resistivity values of the measuring fields can clearly be recognized while the differences between the leakage resistivities are very small. Under real conditions effects like fluctuations in temperature or humidity could even overlay these. This results from the fact that the regression curve of moisture content versus resistivity follows a logarithmic relationship. logarithmically. There is just a small spectrum in which values representing a wet condition vary however a large spectrum exists of possible values representing a dry condition. As a consequence for the application on a bridge deck, it is necessary that a mortar demonstrating high resistivity values in a dry state is used and that the mortar will be given some time to dry out before a reliable detection of leakages is possible. On the other hand even small amounts of water which permeate into the interlayer can be detected under these circumstances.

3.2 Influence of the Leakage Depth

Besides the size of the defective membrane and the resistivity of the mortar under the defective membrane and the resistivity of the remaining field, other parameters could influence the detectability of leakages. The first parameter considered is the depth up to which water has penetrated into the interlayer. Figure 9 shows resistivity values for a leakage depth of 1–3 cm. The actual measuring field starts at a depth of 1 cm and ends at a depth of 2.275 cm which is the space between the two carbon meshes (here shown by the coloured area). As an example the mortar resistivity of 10,000 Ωm and a leakage resistivity of 150 Ω and 700 Ωm are plotted as function of the leakage depth.

The resistivity decreases with depth of the leakage. Notifiable is the sharp drop in resistivity values when the leakage has nearly reached the lower carbon mesh however without touching it. This is probably due to the fact that in the leakage zone almost a low resistance path between the upper and lower carbon mesh is produced. The zone with higher potentials is not strictly limited to the leakage itself, but expands behind the leakage, as can be seen in Fig. 6. As a consequence resistivity values drop quickly before the second carbon mesh is reached. The curve flattens as soon as the lower carbon mesh is reached. The flattening graph results from the use of a mesh-type anode. If a continuous flat plate anode would have been modelled the graph would show constant values after reaching the second carbon mesh. Here, because of the open mesh structure of the anode, the flow lines of the current partly reach the back side of the mesh and lead to slightly decreasing resistivity values behind the extend of the second carbon mesh. For usage on a bridge deck this means that water which seeps under a membrane but does not penetrate sufficiently deep to enter the

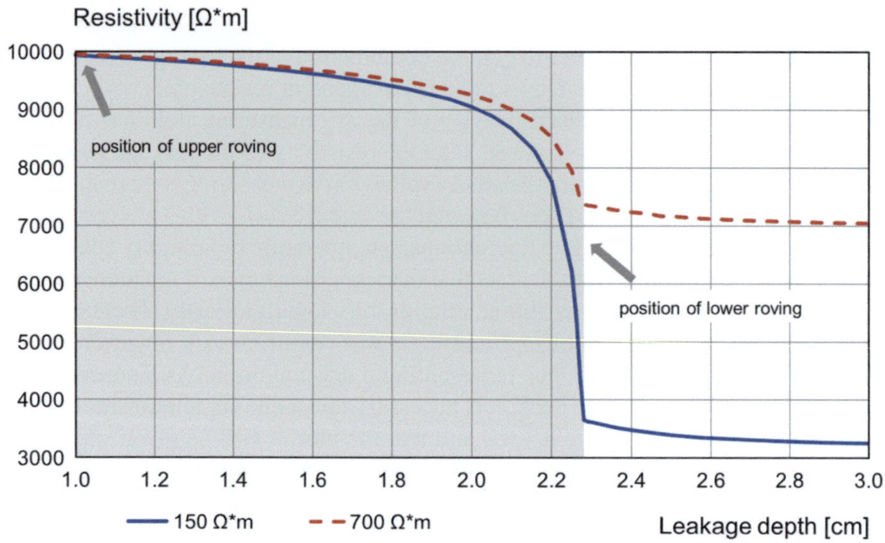

Fig. 9 Influence of the leakage depth based on calculated results obtained from numerical simulations

mortar, will be hard to detect. However, it can be assumed that if the mortar is dried out the leakage water will be absorbed soon and a moisture penetration to greater depth seems likely.

3.3 *Variation of Further Parameters*

In the next steps the remaining parameters shown in Table 1 are investigated. It became obvious that the shape of the leakage, the position of the leakage in the measuring field and the quantity of leakages (compared to a reference leakage with the accumulated size of the variety of smaller leakages) just have a negligible influence on the detectability. This is exemplified in Fig. 10 for a dry mortar resistivity of 5000 Ωm and the case that the leakage depth is at a maximum of 3 cm, i.e. 5 mm beyond the position of the second roving.

The differences in resistivity for the smaller leakage sizes result from a different amount of anode material in the leakage zone. Under the condition of very small leakages, the shape of the leakage and the mesh structure of the anode lead to relatively high differences in the amount of anode material touched by the leakage and therefore to different resistivities. At higher leakage sizes this impact becomes smaller.

Only the geometry of the anode affected the drop in resistance values, but merely if the leakage depth has not already reached the second carbon mesh. Figure 11

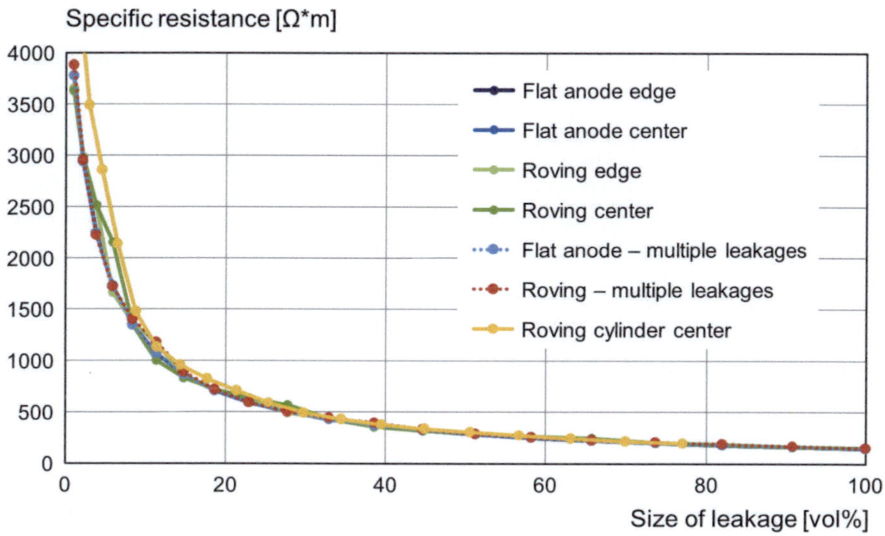

Fig. 10 Parameters with negligible impact on the detectability of leakages

shows this effect. In contrast to Fig. 10 where all parameters were investigated under the condition that the leakage depth covers the entire measuring depth up to the second carbon electrode, Fig. 11 shows a leakage depth of 2 cm, however such a short circuit cannot be assumed. In this case the geometry/geometrical arrangement

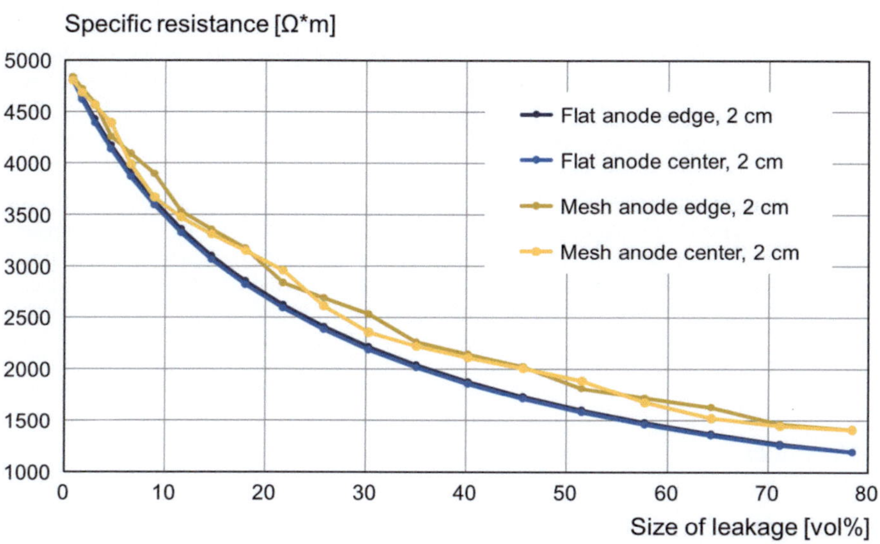

Fig. 11 Impact of the anode structure in combination with the leakage depth

of the anode has an effect on the drop in resistivity. While flat anodes lead to low spreading resistances, more discrete anodes generally result into higher resistances. As a consequence, the mesh anode will lead to lower drops in resistivity values [8].

However, small over- or underestimates of resistivity drops are not critical for the detection of leakages. There will always be a variety of measuring fields in close proximity, so the changes in resistance values can be seen relative to the neighbouring measuring fields. Thus it is possible to adjust if an actual leakage occurs or if other boundary conditions lead to the measured change in values.

4 Conclusions and Outlook

Based on the findings of the numerical simulations the following conclusions can be drawn:

- Leakages having a limited size of a few percentages of the total surface area are detectable
- Resistivity values of a mortar located under an intact membrane in a dry environment must be high compared to the resistivity values obtained for a water saturated condition which means that leakages can be detected as soon as some desiccation of the mortar proceeded.
- The main impacts of various boundary conditions on the detectability of leakages were demonstrated

This paper focused on the theoretical aspects regarding detection of leakages. Nevertheless, as mentioned before, under real conditions parameters like temperature and humidity have an additional effect on the actual detectability of leakages. Therefore additional laboratory tests will be carried out under different climatic conditions and with temperature compensation. Furthermore, the theoretical results will be validated under practical conditions on a demonstration panel with a size of about 100 m^2 by artificially produced leakages.

Acknowledgements The authors would like to thank the BMBF for sponsoring the project and the project executing organization, "VDI Technologiezentrum GmbH (VDI TZ)" for the support. The project partners are: Bundesanstalt für Straßenwesen (BASt), Eurovia Beton GmbH NL Bauwerksinstandsetzung (Projektkoordinator), FTA Forschungsgesellschaft für Textiltechnik Albstadt mbH, Massenberg GmbH, instakorr GmbH (Darmstadt) and Sto Cretec GmbH, Tochtergesellschaft der Sto SE & Co. KGaA.

References

1. Raupach M, Gulikers J, Reichling K (2013) Condition survey with embedded sensors regarding reinforcement corrosion: Bauwerksüberwachung mit eingebetteten Sensoren hinsichtlich der Korrosion von Stahl in Beton. In Materials and corrosion, vol 64. Nr 2, pp 141–146. ISSN 1521-4176

2. Raupach M, Dauberschmidt C, Wolff L (2006) Monitoring the moisture distribution in concrete structures. In: Alexander MG, Beushausen H-D, Dehn F, Moyo P (eds) Proceedings of the International Conference on Concrete Repair, Rehabilitation and Retrofitting (ICCRRR), Cape Town, South Africa, 21–23 November, Taylor & Francis Group, London, pp 166–167
3. Raupach M, Dauberschmidt C, Warkus J (2003) Indirekte Bestimmung der Feuchteverteilung in zementösen Baustoffen durch Elektrolytwiderstandsmessungen (in German). In: Innovative Feuchtemessung in Forschung und Praxis, Bauwesen-Geotechnik- Umwelttechnik, Kolloquium mit Workshop, 3./4. Juli 2003. SMG Soil Moisture Group der Universität Karlsruhe, Karlsruhe
4. Driessen C, Raupach M (2016) Intelligent, multifunctional textile reinforced concrete interlayer for bridges. In: Dehn F, Beushausen H-D, Alexander MG, Moyo P (eds) Proceedings of the 4th International Conference on Concrete Repair, Rehabilitation and Retrofitting (ICCRRR), Leipzig, 5–7 October 2015. CRP Press Taylor & Francis Group, London [u.a]. ISBN 978-1-138-02843-2
5. Moosbrugger C. Electrical and magnetic properties of metals. ASM International. Materials Properties Database Committee. ISBN: 0-87170-690-3
6. Harnisch J (2003) Untersuchungen zum nachträglichen Einbau von Multiring-Elektroden im Beton: Investigations on Subsequently Embedded Multi-Ring-Electrodes in Concrete. Aachen, Technische Hochschule, Fachbereich 3, Institut für Bauforschung, Diplomarbeit
7. Brameshuber W, Raupach M, Schröder P, Dauberschmidt C (2003) Non-destructive determination of the water content in the concrete cover using the multi-ring-electrode. In: International Symposium Non-Destructive Testing in Civil Engineering (NDT-CE), Berlin, September 16–19, 2003, Poster No. 41. Deutsche Gesellschaft für Zerstörungsfreie Prüfung, DGZfP, Berlin
8. Raupach M (1992) Zur chloridinduzierten Makroelementkorrosion von Stahl in Beton. In: Schriftenreihe des deutschen Ausschusses für Stahlbeton. Beuth, Berlin, pp 16–17

Numerical and Experimental Development of Gradient Potential Measurement for Corrosion Detection in Reinforced Concrete

S. Garcia and F. Deby

Abstract Half-cell potential mapping is commonly used to detect corrosion risks in Civil Engineering structures. This method uses a reference electrode positioned on the surface of concrete for measuring potential difference by using a voltmeter connected to the reinforcement bar. However, there are two major drawbacks in the implementation of this method: the electrical connection to the reinforcement and its electrical continuity. In order to overcome these disadvantages, a new method is proposed. Instead of using electrical connection to rebar, a second reference electrode is used, also positioned on the surface. These two electrodes configuration result in electrical potential gradient on the concrete surface. By performing both experimental work and numerical modelling, this configuration is tested. For experimental work, two reinforced concrete slabs ($3 \times 3 \times 0.15$ m) were casted to be close to actual structural conditions. Then corroded areas with controlled sizes are created through a specific device allowing chloride migration. The thickness of the concrete cover has an influence on the measures, which is why the first slab includes rebar where the cover is fixed, while it is variable in the second slab. For modelling, a parametric study using a finite element model is performed. This model allows the study of the influence of several parameters such as resistivity, corroded area size, concrete cover, and also the measurement process. It is the analysis of the corrosion current, as well as potential mapping that allows the correlation between the experimental and modelling results. Discussion of both results concludes feasibility of this method and confirms its benefits compared to usual half-cell potential mapping.

Keywords Reinforced concrete · Corrosion · Modelling
Potential · Non-destructive testing method

S. Garcia (✉) · F. Deby
LMDC, Université de Toulouse, INSA, UPS, Toulouse, France
e-mail: sylvain.garcia.bec@gmail.com

C. Andrade et al. (eds.), *Service Life and Durability of Reinforced Concrete Structures*, RILEM Bookseries 17, https://doi.org/10.1007/978-3-319-90236-4_6

1 Introduction

The main factor of reinforced concrete building destruction is the corrosion of reinforcement bars. It is possible to observe it in a world-wide scale because of structures ageing. That is why detection and diagnostic of reinforcement corrosion are a major challenge in order to extend the structure durability. Non Destructive Techniques (NDT) are useful in risk area detections: they give quick and less expensive information compared to intrusive methods. If the diagnosis is carried out early and regularly the cost of structure maintenance decreases, therefore the development of methodologies to detect and assess reinforcement corrosion is crucial [1–3].

This study is realised in the framework of the ATALANTE project, whose the goal is to develop a robot for non-destructive testing (NDT) of horizontal concrete surfaces. One of the objectives is to detect reinforcement corrosion as quickly and simply as possible.

So this study aims to develop a NDT method with high measuring speed and which does not requires an electrical connection. Usual half-cell potential mapping [4] and linear polarization resistance [5] are commonly used for the corrosion detection but do not suit to requirements because of the electrical connection. This connection needs specific materials and time, without this step, both auscultation time and cost will be reduced. This study aims to develop a NDT method which does not need this electrical connection. Therefore, a potential gradient method has been chosen using two reference electrodes [6].

The development of this methodology of measurement has been performed by two ways, firstly a numerical study and secondly a full-scale testing on large reinforced concrete slabs.

The numerical study was achieved to understand the influence of resistivity and concrete cover on the potential corrosion. The advantage of the numerical models is to easily provide a large quantity of information which allows performing a parametric study of all the influent parameters. At the end it is possible to better define the experimental tests.

Before designing the experimental tests, an accelerated localized corrosion protocol on a small scale is proposed to develop corrosion spots on large-scale slabs (3 m × 3 m × 0.15 m). Then the methodology of measurement is achieved on the slab and it is shown that the detection of localised corrosion is possible by the assessment of potential gradient without connection to the rebar.

The exploitation of the results and the comparison with recommendations allow concluding on the relevance of the proposed methodology.

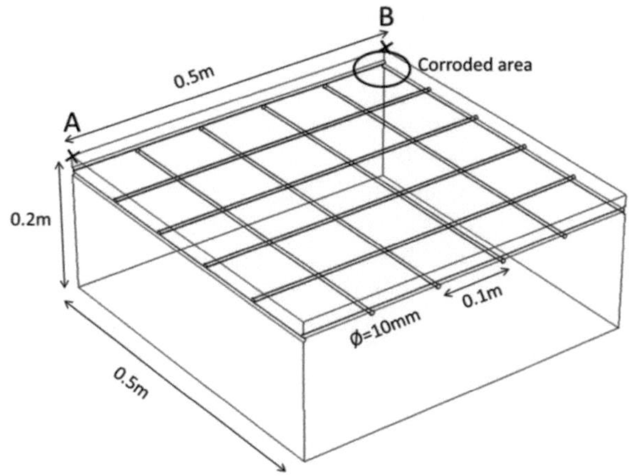

Fig. 1 Prism geometry in reinforced concrete for the modeling of the streamlines distribution

2 Modelling

2.1 Model Description

Finite Element Methods (FEM) are performed using the module: "DC Conductive Media" from COMSOL Multiphysics®. Numerical simulations are achieved on square prisms of 1m × 1m × 20 cm concrete for a 10 mm rebar diameter and a horizontal spacing of 10 cm. It is possible to perform modeling on a quarter of the prism (Fig. 1) through the two axial symmetries.

In order to determine numerically the current densities in the model, it is necessary to determine the electrical potential at any point of the geometry. Ohm's law links the local electric current density to the conductivity (or resistivity) and the potential gradient (Eqs. 1 and 2) under the condition of charge conservation (Eq. 3).

$$j = -\sigma \cdot \nabla\phi_s \quad (1)$$

$$j = -\frac{1}{\rho} \cdot \nabla\phi_s \quad (2)$$

$$\nabla \cdot j = 0 \quad (3)$$

Our simulations take place in a secondary current distribution between corroded and non-corroded bars, forming a galvanic cell in the system.

Electrochemical kinetics are described with Butler-Volmer equations; they are used as boundary conditions, corroded zone being the anode (Eq. 4) and non-corroded the cathode (Eq. 5).

$$j_a = j_{a\,corr} \times \left[\exp\left(\log(10) \times \frac{(V - E_{corr\,a})}{\beta_{a\,a}}\right) - \exp\left(-\log(10) \times \frac{(V - E_{corr\,a})}{\beta_{c\,a}}\right)\right] \quad (4)$$

Table 1 Electrokinetics parameters

		Anodic	Cathodic		
j_{corr}	Corrosion current density	5.0×10^{-3}	A/m^2	2.5×10^{-4}	A/m^2
β_c	Cathodic Tafel coefficient	0.15	V/dec	0.15	V/dec
β_a	Anodic Tafel coefficient	0.09	V/dec	0.4	V/dec
E_{corr}	Corrosion potential	−0.65	V/ECS	−0.15	V/ECS

$$j_c = j_{c\ corr} \times \left[\exp\left(\log(10) \times \frac{(V - E_{corr\ c})}{\beta_{a\ c}}\right) - \exp\left(-\log(10) \times \frac{(V - E_{corr\ c})}{\beta_{c\ c}}\right)\right] \quad (5)$$

Parameters used for the simulation were measured by Sohail et al. [7] on a carbonated concrete with properties very close to the ones of the concrete used in this study (Table 1).

The numerical study involves the combination of resistivity and concrete cover which are the most relevant ones regarding potential measurement.

Different values of resistivity (ρ) are considered, 200, 475, 750, and 1300 Ω m, and regarding concrete cover (h), the considered values are 2, 4 and 5 cm.

2.2 *Effect of Resistivity*

Resistivity of concrete slab combines numerous physical parameters like: porosity, humidity, and ionic concentrations of the pore solution; resistivity is consequently a crucial factor for the corrosion in concrete.

Figure 2 identifies the influence of resistivity on the mapping of electrical potential obtained at the surface. For low resistivity 200 Ω m a range from −190 mV/SCE for corroded zone to −300 mV/SCE above the corroded zone is observed (Fig. 2a). For high resistivity this scale is between −160 and −290 mV/SCE (Fig. 2d). There is an overall decrease in range potential with the decrease of resistivity (Fig. 2a–d). In addition, there is a concentration of highest electro-negative values (<−0.25 V/SCE) at the center of the corroded area for a high resistivity, where these values are more diffuse with the lowest resistivity. We can conclude that the increase of ohmic drop due to high resistivity induces current lines containment near the corroded area.

Figure 3 shows the local corrosion current densities along the upper part of the corroded reinforcement. In the center of the corroded area (x = 0.5 m), the local current density is higher for low resistivity (in absolute value): 2.78×10^{-2}A/m^2 for 1300 Ω m and 13.7×10^{-2}A/m^2 for 200 Ω m. However, on the non-corroded

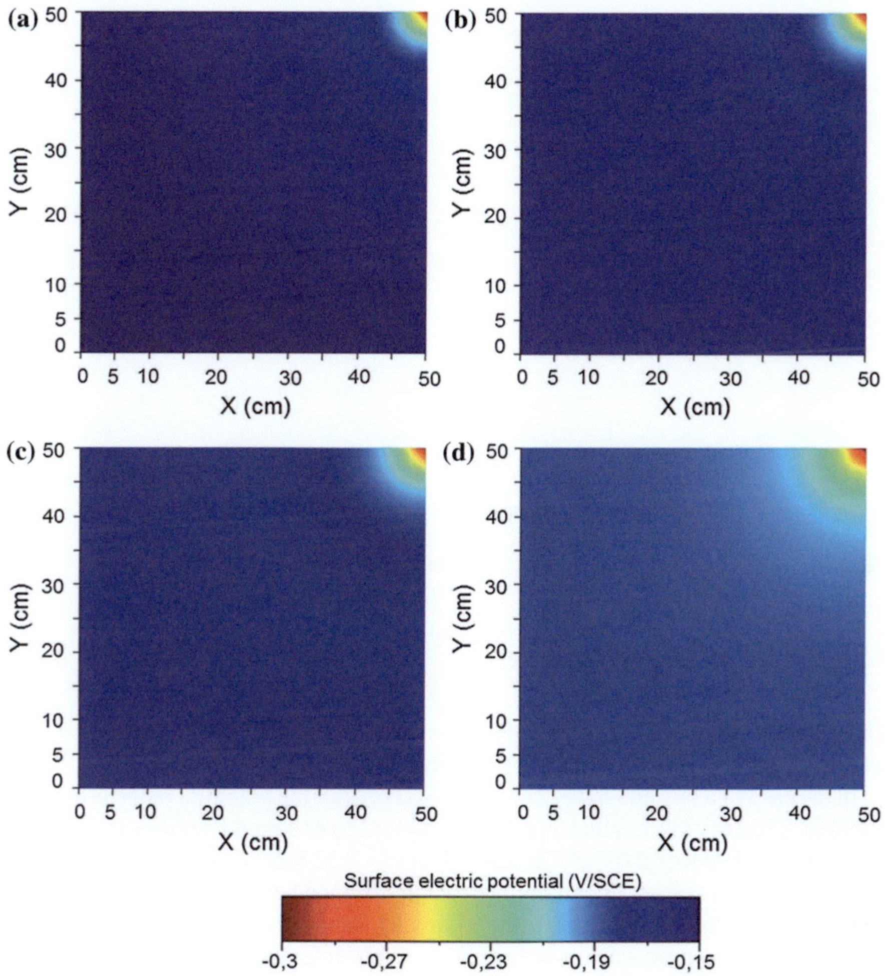

Fig. 2 Half-cell potential mapping in function of resistivity: **a** ρ = 200 Ω m, **b** ρ = 475 Ω m, **c** ρ = 750 Ω m, **d** ρ = 1300 Ω m; corroded length = 1 cm, h = 2 cm

side, the maximum current density is about 2.35×10^{-2}A/m^2 for 1300 Ω m at x = 0.495 m and 1.20×10^{-2}A/m^2 for 200 Ω m. This validates the assumption that the concentration of streamlines is significant near the frontier between corroded and non-corroded areas.

According to this modelling before the half-cell potential measurements, it would be preferable to not over-humidify the studied surface.

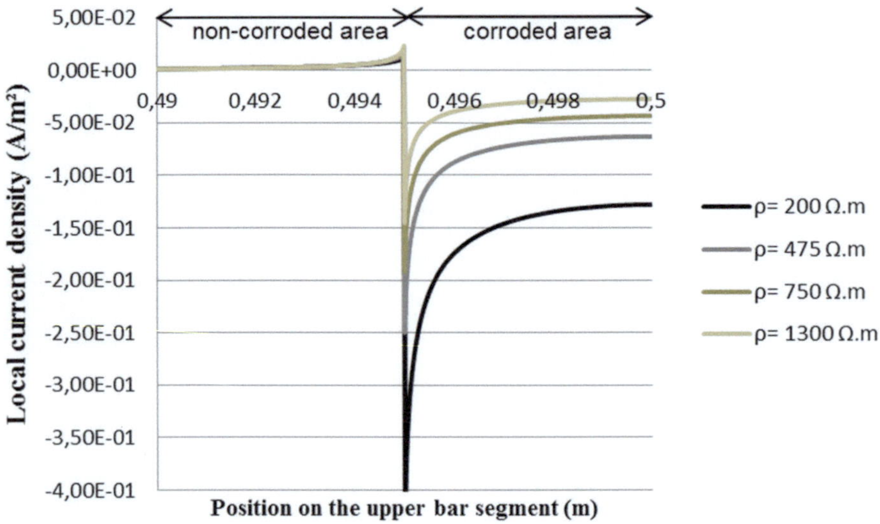

Fig. 3 Current density near the frontier of corroded zone for four resistivities

2.3 *Effect of Concrete Cover*

The distance from the reinforcement to the surface of the concrete has a major impact on the potential values obtained at the surface.

In the case of a low concrete cover of 2 cm (Fig. 4a), the potential range is between −190 to −300 mV/SCE which is greater than for larger concrete cover of 4 and 5 cm (Fig. 4b, c) −190 to −240 V/SCE and −190 to −230 V/SCE respectively.

These differences are due to the increasing distance between rebar and measurement zone and thus a removal of the more electronegative iso-potentials from the surface of the concrete slab. A strong concrete cover generates a contrast reduction similar to a loss of information. Elsener et al. [8] have shown experimentally the same results on a smaller slab with a rebar length of 30 cm, a corrosion area of 0.5 cm and a resistivity of 1300 Ω m.

Figure 5 shows the global current versus resistivity for these three concrete covers. For a given resistivity, the global current increases with the concrete cover: at 200 Ω m it is about 2.21×10^{-5} A for h = 2 cm against 2.28×10^{-5} A for h = 4 cm. When the concrete cover is doubled (from 2 to 4 cm at 200 Ω m), there is an increase of the global current of 3.4 and 4.1% for 5 cm of thickness. For a resistivity of 1300 Ω m, the increase is only about 2.4% for 4 cm of concrete cover and 2.9% for 5 cm. Looking at the lower side of the reinforcement, there is no influence of the concrete cover on the local current density along the rebar. This means that when the concrete cover is low, the streamlines between rebar and the surface of the slab are confined, giving a lower current density of the upper side of the reinforcement, and this phenomenon is even more important when the resistivity is low. This influence is observed because, when

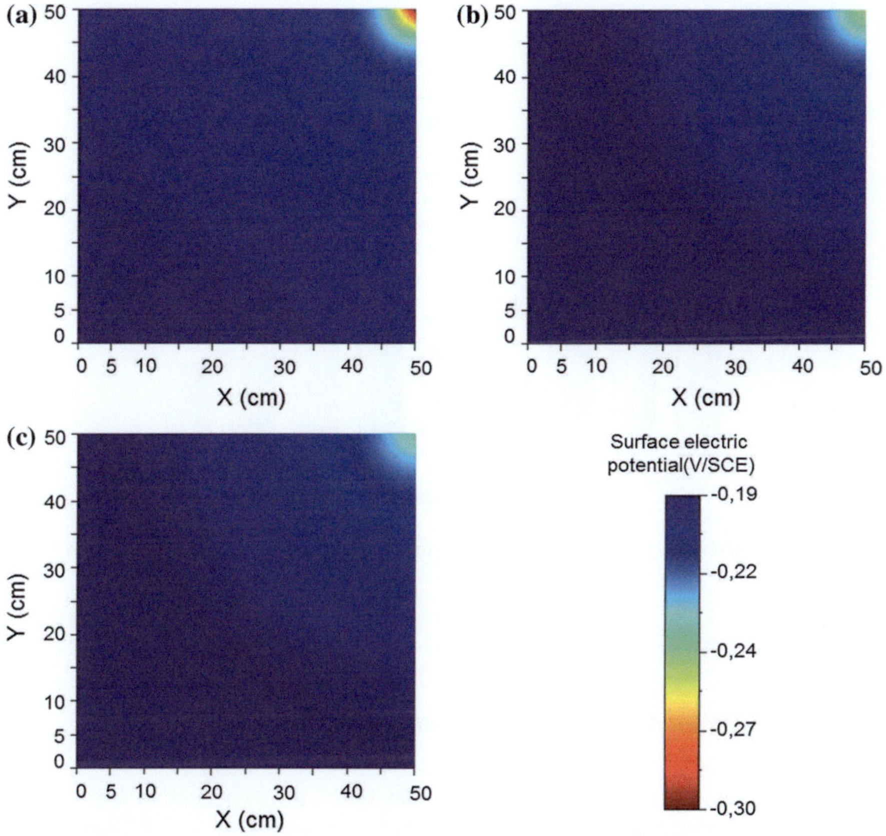

Fig. 4 Half-cell potential mapping in function of concrete cover: **a** h = 2 cm, **b** h = 4 cm, **c** h = 5 cm; corroded length = 1 cm, ρ = 200 Ω m

the resistivity is low, the current lines will polarize the steel over greater distances. This is why at 200 Ω m the overall global current for a concrete cover of 5 and 2 cm is larger than for 1300 Ω m.

To conclude about the concrete cover, when it is increasing, current densities are increasing too, but there is a loss of information at the surface of the concrete giving a smaller potential gradient.

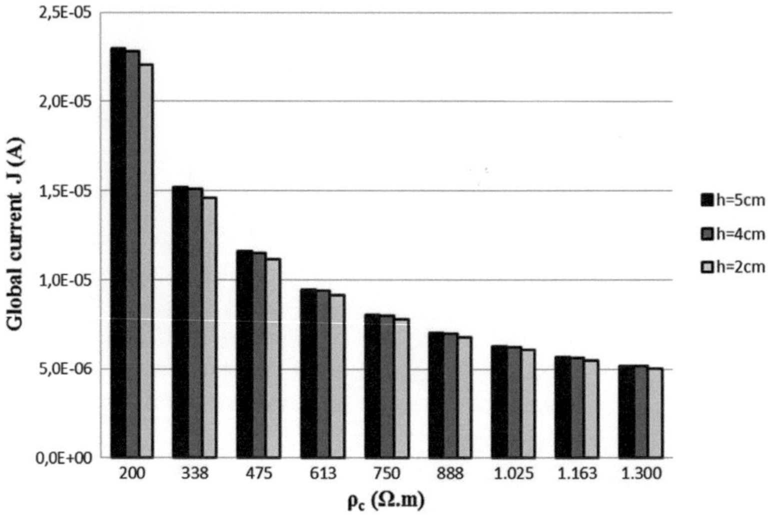

Fig. 5 Global current in function of the resistivity for three concrete cover; corroded length = 1 cm

3 Experimental Testing

3.1 Accelerated Corrosion Method

In order to develop the method of corrosion detection, it was necessary to be able to create quickly controlled corroded areas. The chosen process was a chloride migration method [9, 10].

A hollow cylinder is glued to the surface of the reinforced concrete slab above a rebar intersection (Fig. 6). The cylinder is then fills with 300 mL of NaCl solution at a concentration of 20 g/L. Rebar acts as working electrode while counter electrode is made of stainless steel. Reference electrode and counter-electrode are positioned inside the cylinder. A generator is set up to connect the plates and counter-electrode in order to output a DC voltage of 20 V.

This test is performed on a thin slab of 25 × 22 × 7 cm dimension (sample B); the reinforcement is composed of reinforcing mesh of three bars spaced out of 10 cm. In parallel a second sample (sample A), will be tested in the same conditions without the current injection, being the control sample.

After 70 h of DC voltage, sample A and B were cut in order to measure the chloride penetration depth by spraying a 0.1 M silver nitrate solution on to the split section. The formation of a white precipitate composed of silver chloride indicates the presence of chlorides (Fig. 7). Specimen A has a maximum chloride penetration depth of 2.2 cm, which is only due to diffusion, but also to capillary absorption. Specimen B has a way greater penetration depth, up to 5.9 cm. This higher result is of course due to ions migration.

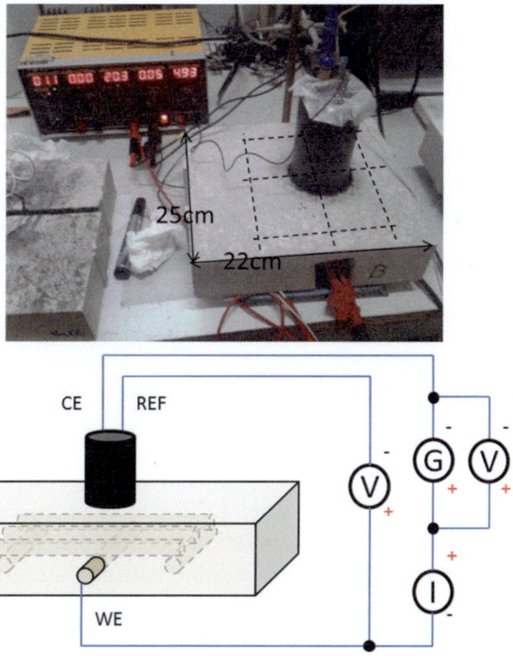

Fig. 6 Picture and schema of accelerated corrosion process

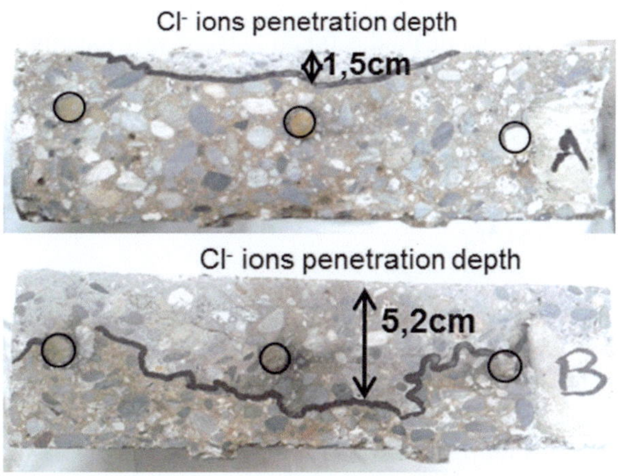

Fig. 7 Pictures of sample A (diffusion/absorption) and B (migration/diffusion/absorption) after the test, split and $AgNO_3$ spraying

At the beginning of the experiment, the chloride/hydroxide ions concentration ratio tends to the infinite because the solution was neutral. But during the test the pH increases to 13, inducing a reduction of the ratio to a value of 0.58 at the end of the

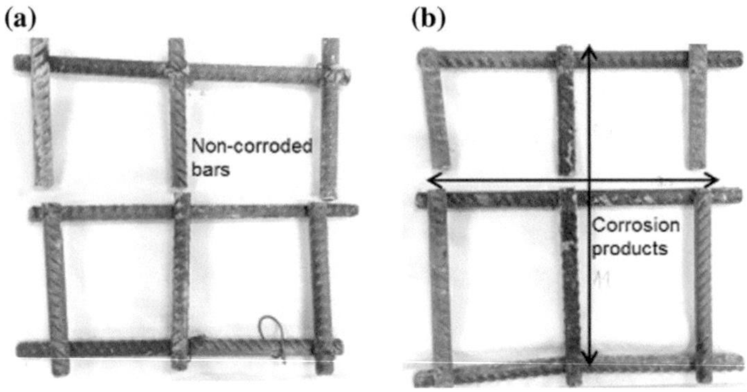

Fig. 8 Pictures of sample A (diffusion/absorption) and B (migration/diffusion/absorption) extracted reinforcement

test. This means that hydroxide ions are 1.7 times more injected into the slab than chloride ions, the efficiency of the operation is no longer interesting. For a deeper chloride migration, it would have been required to renew the solution of the cylinder.

The autopsy of specimens shows that the sample A contains no trace of corrosion (Fig. 8a) on the contrary for specimen B (Fig. 8b) severe corrosion is observed on the central reinforcement. The upper bar is corroded along 20 cm on its upper side and of 14.5 cm on the lower side; the lower bar is corroded on its upper part on 19.5 and 13.5 cm on its lower part.

The accelerated corrosion method is functional and provides significant results in a relatively short period of time (for this concretes formulation). This method can now be used on a larger scale to create sites of localized corrosion.

3.2 Structural Testing

Description of large-scale testing. Following the development of the accelerated corrosion method, large-scale testing was possible. A reinforced concrete slab of $3 \times 3 \times 0.15$ m was casted with a concrete cover of 2 cm and a 10 cm reinforcement horizontal spacing (Fig. 9).

Ten months after casting, a first corrosion site was created with a diameter of about 10 cm. It is possible to carry out measurement with the two reference electrodes method without electrical connection to reinforcement.

Measurement method. As said before, a potential gradient method is chosen using two reference electrodes to achieve a mapping of the concrete slab. The copper-copper sulphate electrode is selected because of its advantages: a simple maintenance (maintain the copper sulphate solution saturated) and an easy and quick measurement using a wheel electrode instead of traditional stick electrode. First of all, the step and the distance between electrodes were chosen equal to the size of the corroded zone:

Fig. 9 Picture about 3 × 3 × 0.15 m reinforced concrete slab before casting

10 cm. Figure 10 shows in green the first scan required to build a potential mapping with two reference electrodes. This first scan needs that the electrodes are both in the measurement axis. Once the first measure is achieved ($V_{(0,1)}$), the device is moving in the same axis so, for the second measure ($V_{(0,2)}$), the first electrode is located in the past position of the second and so on ($V_{(0,m)}$).

The second passage ($V_{(1,m)}$) is performed in the same axis as the first, but the axis of the electrodes is perpendicular to the measures. Others ($V_{(n,m)}$) are the same as for the second passage but it is necessary to move the device of 10 cm (the electrodes distance) in the axis of electrodes.

This method does not allow the building of a classical half-cell potential mapping reporting to SCE. Nevertheless, this method allows to build two kinds of mapping: a gradient potential mapping (with the raw data) and an equivalent potential mapping.

Fig. 10 Schema of the mesh measurements for the potential mapping method, using two reference electrodes

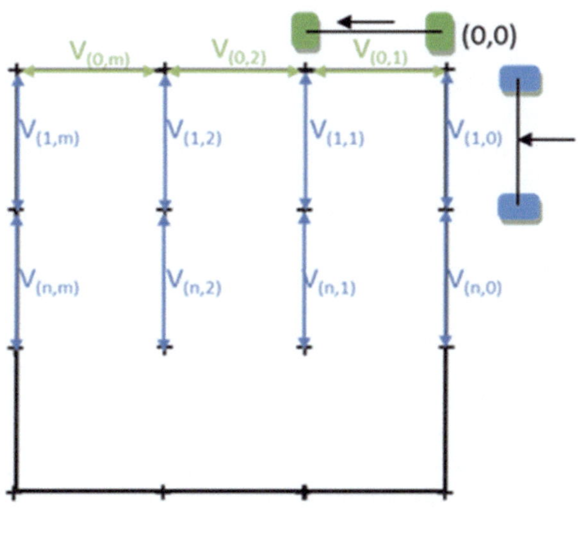

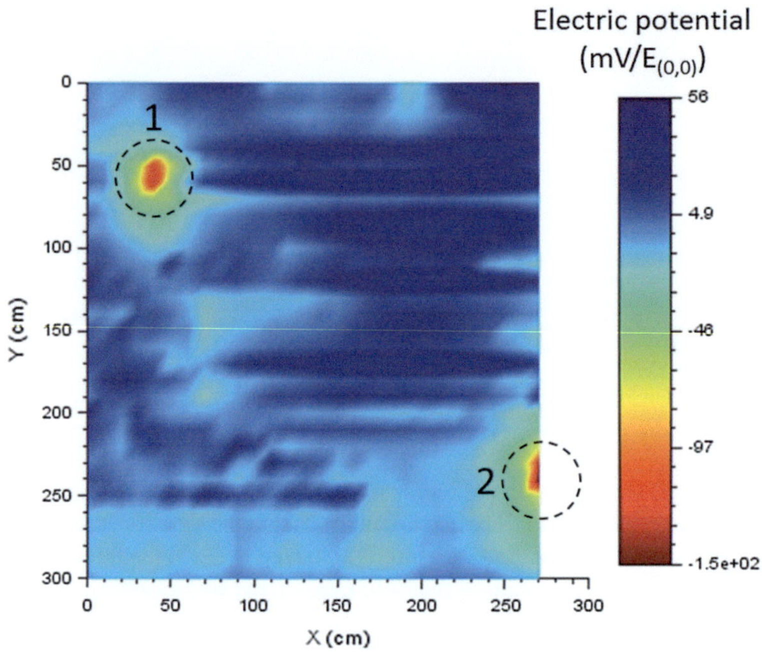

Fig. 11 Equivalent potential mapping using two reference electrodes with a measuring step of 10 cm, giving the detection of two corroded areas; $E_{(0,0)} = 0$ V

About the latter, a reference is needed and, in this study, this is a measurement point with the (0,0) coordinates whose the value is taken arbitrary equal to 0 V. This article will only presents the results obtained with the equivalent potential mapping.

Results. The use of the device on the slab surface gives the equivalent potential mapping presented in Fig. 11. The potential at coordinates $x = 0$, $y = 0$ is arbitrarily chosen to 0 V as previously mentioned. On this mapping we can see a first corroded area, numbered 1, where we used the accelerated corrosion method. The potential reaches -133 V/$E_{(0,0)}$ whereas distant from 30 cm the potential is about 90 to 140 V higher. We can also detect a second corroded area with a potential reaching -148 V/$E_{(0,0)}$. This corrosion is not due to the accelerated method but to a rebar deformation. This deformation causes a local rise of rebar and therefore a very low concrete cover and a local carbonated corrosion.

Figure 12 is a half-cell potential mapping carried out around the area 1 (Fig. 11) with the more electronegative value as the center of mapping. This mapping is build using a saturated calomel electrode, to compare previous results with the ASTM C876-91 (Table 2) conclusions. The step measurement is about 1 cm which is ten times finer than the previous system but in a smaller area (30×40 cm).

According to ASTM C876-91 (Table 2), such as the potentials within the circle are between -125 and -250 mV/SCE, the corrosion risk is intermediate (probability about 50%). But looking at the isopotentials which are of a typical circular shape and

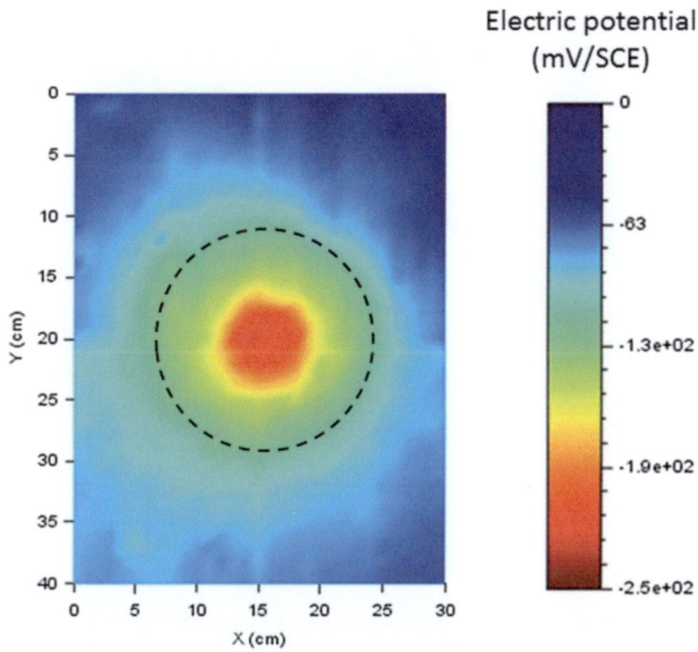

Fig. 12 Half-cell potential mapping around zone 1 using saturated calomel electrode with a measuring step of 1 cm

Table 2 ASTM C876-91 [11]

E_{corr} (mV/SCE)	Corrosion risk
$E_{corr} < -425$	Severe corrosion
$E_{corr} < -275$	High (P>90%)
$-275 < E_{corr} < -125$	Intermediate (P = 50%)
$E_{corr} > -125$	Low (P < 10%)

the potential gradient of 250 mV by 20 cm, we can claim that these bars are corroded in this area. Therefore, it is more interesting to look at the potential gradients rather than talking in terms of "absolute value" and corrosion risk.

The build of this equivalent potential mapping shows the feasibility of corrosion detection from the concrete surface without drilling and connecting the device to the rebar and therefore no requirement to check rebar continuity. These advantages reduce the number of manipulations which could facilitate automate measures.

Comparison to modeling. Other half-cell potential measurements were carried out, from the center of the corroded zone 1 to the limit of this zone following the 0y axe. Figure 13 shows this measurement for three different environmental conditions:

- Just after a full day of rain (11th February very wet),
- 3 h after the rain has stopped (11th February wet),

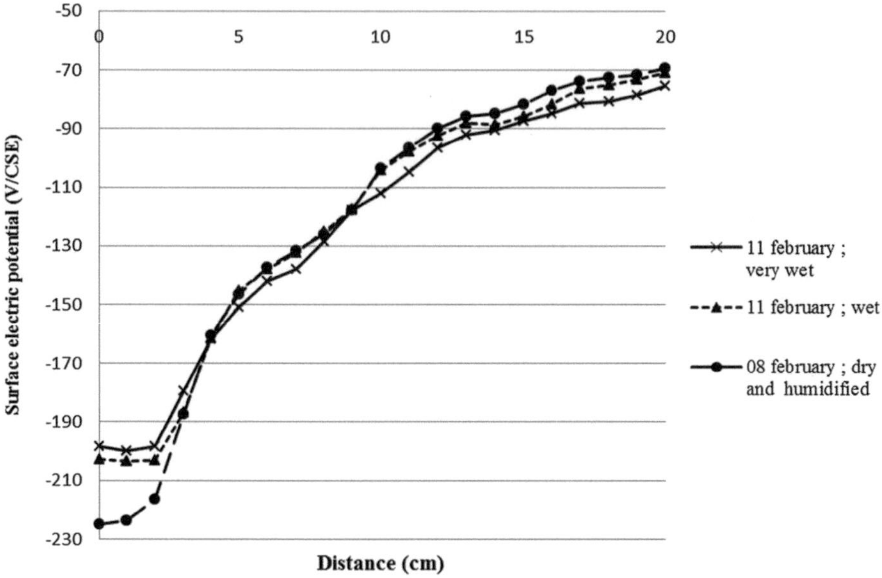

Fig. 13 Half-cell measurement from the center of zone 1 in 0y direction using SCE

– 3 days earlier with a dry weather and a surface wetting of the slab (8th February dry).

These three measuring conditions corresponds to lowest resistivity for very wet and to highest resistivity for dry. The very wet condition gives a most negative value of −200 mV and a gradient potential of 129 mV; for the wet condition the most negative value is about −204 mV and a gradient potential of 133 mV; and for the dry condition the most negative value is about −225 mV with a gradient potential of 150 mV. We can see that the higher the resistivity is, more negative the potential above the corroded zone are. And conversely, far from the corroded zone, the higher the resistivity is, less negative potentials. Humidity is a determining factor on the values of potential but also on the observed potential gradients.

Figure 14 is an electric potential plot on [AB] (Fig. 1) resulting of modelling. The trends are identical to those obtained experimentally: for a resistivity of 200 Ω m, potential gradients are the lowest (130 mV) while for 1300 Ω m, potential gradients are the highest (230 mV).

About resistivity it can be concluded that a significant humidification is not the best choice to make a good identification of the corrosion sites. Moreover, for high resistivity, current densities and therefore global current due to corrosion are lower, and finally potential values on concrete surface are more electronegative. However, ASTM highlights that more negative the potential is, more important the corrosion is, which is not necessarily true.

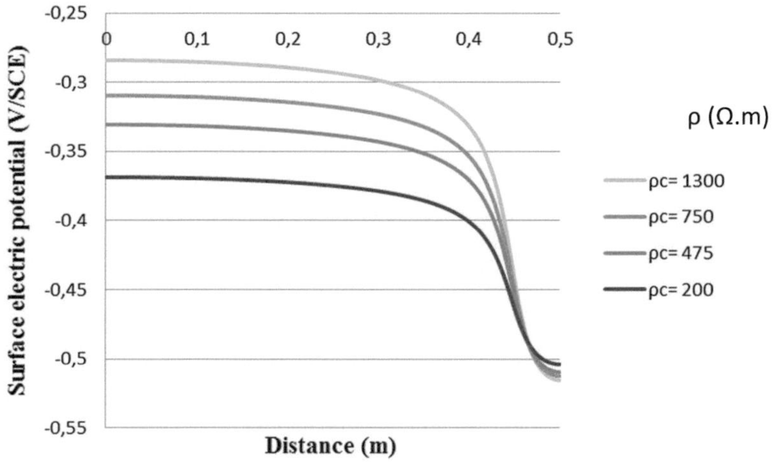

Fig. 14 Electric potential plot on [AB] (at the surface of the slab) h = 2 cm; corroded length = 10 cm

4 Conclusion

In this paper a new method is proposed to measure corrosion potential gradient of concrete reinforcement. The principle consists in measuring the potential gradient between two potential electrodes without connection to the reinforcement. This new approach is interesting because it does not require drilling concrete, in order to have the electrical connection, and to check the continuity of reinforcement.

The feasibility of the proposed methodology is firstly tested by means of numerical modelling by using finite element model. The model is based on Butler-Volmer equations to describe the corrosion kinetics of steel and involves both resistivity and concrete cover.

Modelling allows visualization of mechanisms and gives a good comparison to experimental results. At this point, the aim of modeling is not to give quantitative values but it would be possible to have a quantitative comparison of the corrosion mechanism with a set of Butler-Volmer parameters from the studied concrete formulation.

An accelerated corrosion method has been developed to create localized corrosion of 10 cm diameter inside a big scale slab ($3 \times 3 \times 0.15$ m). Thereafter, this defect has been detected using the gradient potential measurement with the bi-electrode method without connection to the rebar.

Usual half-cell mapping are more dependent on environmental conditions such as water saturation and temperature, which may vary from day to day, unlike the observation of potential gradients. The exploitation of result proposed by ASTM is therefore a rather bad indicator of the presence of corrosion contrary to the observations that can be made with the gradients of potential. Moreover bi-electrode mapping offer a less constraining method such as connection to reinforcement is not required.

Acknowledgements The authors want to acknowledge the French Public Bank for Investment (BPI) for the funding of ATALANTE project and the Company ETANDEX.

References

1. Leibbrandt A, Caprari G (2012) Climbing robot for corrosion monitoring of reinforced concrete structure. In: International conference on applied robotics for the power industry 2012, ETH Zurich, Switzerland
2. Reichling K, Raupach M (2014) Method to determine electrochemical potential gradients without reinforcement connection in concrete structures. Cement Concr Compos 47:3–8
3. Raupach M, Reichling K, Wiggenhauser H, Stoppel M, Dobmann G, Kurz J (2009) BETOSCAN—an instrumented mobile robot system for the diagnosis of reinforced concrete floors. In: Proceedings ICCRRR 2008, Kapstadt, South Africa, pp. 24–26
4. Stratfull RF (1959) Progress report on inhibiting the corrosion of steel in a reinforced concrete bridge. Corrosion 15(6):65–68
5. Guliker J (1997) Development of a galvanic monitoring probe to improve service life prediction of reinforced concrete structures with respect to reinforcement corrosion. Constr Build Mater 11(3):143–148
6. François R, Arliguie G, Bardy D (1994) Electrode potential measurements of concrete reinforcement for corrosion evaluation. Cem Concr Res 24(3):401–412
7. Sohail MG, Laurens S, Deby F, Balayssac JP (2013) Significance of macrocell corrosion of reinforcing steel in partially carbonated concrete: numerical and experimental investigation. Mater Struct 48:217–233
8. Elsener B, Andrade C, Gulikers J, Polder R, Raupach M (2003) Half-cell potential measurements—potential mapping on reinforced concrete structures. Mater Struct 36:461–471
9. Ait Mokhtar A, Belarbi R, Benboudjema F, Burlion N, Capra B, Carcassès M, Colliat JB, Cussigh F, Deby F, Jacquemot F, Larrard T, Lataste JF, Le Bescop P, Pierre M, Poyet S, Rougeau P, Rougelot T, Sellier A, Séménadisse J, Torrenti JM, Trabelsi A, Turcry P, Godoy HY (2013) Experimental investigation of the variability of concrete durability properties. Cem Concr Res 45:21–36
10. Gao XX, Deby F, Arliguie G, Lang A, Gourbeyre Y (2014) Création de défauts artificiels sur des âmes en tôles d'éléments en béton, DiagnoBéton 2014. Annales du bâtiment et des travaux publics, Toulouse, France 66(2014):144–148
11. ASTM C 876-91 (1991) Standard test method for half-cell potentials of reinforcing steel in concrete. Annu Book ASTM Stand 4:2

Advances in Characterization of Gas Transport in Concrete: Determination of Oxygen Diffusion Coefficient from Permeability Coefficient and Porosity

P. Linares, C. Andrade and D. Baza

Abstract One of the most relevant mechanisms that influence the life service of concrete (carbonation, compactness, alkaline protection) is gas diffusion through its mass. However, determination of gas diffusion coefficient in concrete is not a simple task. There is not a general standard which determines the test procedure. Other approaches or alternative parameters could be used in order to obtain the gas diffusion coefficient, including the use of permeability coefficient. Both parameters are dependent on material porosity and moisture content. The literature already proposes generic correlations for these parameters, but these correlations do not support the direct derivation of the gas diffusion coefficient. This paper presents the results of research carried out to analyse these generic correlations and to propose specific expressions that support the derivation of a value for the oxygen diffusion coefficient, based on the porosity and permeability coefficients. The research was centred on a experimentation process to obtain these parameters. A diffusion chamber was designed and built for use with two types of concrete mix, two distinct concrete curing processes and three separate values for humidity. Each test case sought to derive values for oxygen diffusion and permeability coefficients.

Keywords Concrete · Diffusion · Oxygen · Permeability

P. Linares (✉) · D. Baza
Eduardo Torroja Institute for Construction Sciences (IETcc), Spanish Research High Council, Serrano Galvache, St, Madrid, Spain
e-mail: plinares@ietcc.csic.es

C. Andrade
CIMNE: International Center for Numerical Methods in Engineering, Universitat Politècnica de Catalunya (UPC), Barcelona, Spain
e-mail: candrade@cimne.upc.edu

C. Andrade et al. (eds.), *Service Life and Durability of Reinforced Concrete Structures*, RILEM Bookseries 17, https://doi.org/10.1007/978-3-319-90236-4_7

1 Introduction

The measurement of the gas diffusion coefficient presents challenges, due to the absence of standard procedures and the requirement to establish adequate air tightness in testing apparatus (referred to here as the diffusion cell).

Another parameter that can characterize and describe the process of gas transport through concrete is the permeability coefficient, which can be measured relatively easily with standardized procedure. The use of this parameter would simplify the characterization of the life service of concrete. Although earlier studies exist, describing correlation among these parameters, there is, as has been observed in the literature, little common ground between the various research efforts. For precisely this reason, it was deemed necessary to carry out further research in this area, as a foundation to the current work, to explore potential relationships among the following parameters: oxygen diffusion coefficient and permeability coefficient.

2 Theoretical Base

Theoretical base is provided on the correlation between diffusion and permeability coefficients.

As it has been mentioned above another parameter that can characterize and describe the process of gas transport through concrete is the permeability coefficient (K), which can be measured relatively easily with standardized procedure.

Some authors identify a relationship between diffusion and permeability coefficients [1].

$$\mathrm{k} = \mathrm{a} \cdot D_{O_2}^{\mathrm{n}} \tag{1}$$

where a : $\mathrm{A}/(8\pi\, D_{O_{2_o}})$, A: sample cross-section, $D_{O_{2_o}}$: diffusion coefficient in air (m^2/s).

Other studies [2–4] experimentally relate both coefficients confirming that this relationship exists, depending on microstructure of pores, as represented by the simplified expression:

$$\mathrm{K} \approx D_{O_2}^{\mathrm{n}'} \tag{2}$$

In each particular study a different value for n and n' was established. These proposed models do not support the direct derivation of the gas diffusion coefficient because they require the diffusion coefficient to be measured to be able to establish n and n′. Therefore a new model is required that allows the determination of a value for n and n′ in a more general way.

Table 1 Concrete types

Concrete	Cement (kg/m^3)	Cement type	Water (kg/m^3)	W/C ratio	Gravel (4/11) (kg/m^3)	Sand (0/4) (kg/m^3)	Additive* (kg/m^3)
1	300	CEMI 42.5 R	165	0.55	928	994	0
2	400	CEMI 42.5 R	166	0.39	886	957	2

*Rebouilt 1000

3 Methods and Materials

The proposed method is based on experimentation to obtain values for the permeability and diffusion coefficients of oxygen of several concrete samples of varied composition, humidity content, w/c ratios and curing processes; with the aim of identifying general relationships between these parameters that might then be used for the analytic estimation of the oxygen diffusion coefficient from the permeability coefficient. Porosity ε was also obtained, being perceived as the most important inherent parameter of concrete related to gas transport. Tortuous porosity, defined as the parameter that relates porosity and tortuosity, has been studied as well, but the results are similar and more complex to obtain, so have not been described in this paper.

Parameters have been measured using the standardized procedures or purpose-built in the laboratory. The oxygen permeability coefficient was experimentally obtained following the method described in EN 83981 [5] and the diffusion coefficient using the purpose-made diffusion cell and methodology. Porosity was obtained with Mercury Injection Porosimetry (MIP) according to ASTM-D4404-10 [6]. Tortuous porosity was obtained experimentally using the following relations:

- Archie´s law: $\rho_e = \rho_0 \cdot \varepsilon^{-\tau}$, where ρ_0 is electrical resistivity of water.
- A relation derived from Archie's law: $\varnothing = \varnothing_0 \cdot \varepsilon^{-\tau}$ (Andrade et al. [7]) where $\varnothing_0$ is the minimum pore diameter in MIP and Ø can be threshold pore diameter $\varnothing_{th}$ or average pore diameter $\varnothing_m$.

3.1 Study Cases

Two concrete types are studied, both using Ordinary Portland Cement but varying the mix and w/c ratio, aspects which have determinant influence on concrete porosity. These concrete types were chosen to represent the range of concrete types regularly used in construction in Spain (Table 1).

For each concrete type two kinds of curing process were studied, due to the fact that this is a further aspect that has a strong influence on porosity:

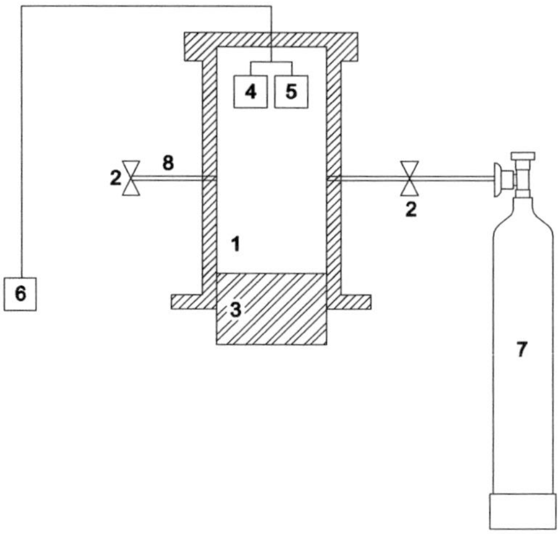

Fig. 1 Measurement equipment

- Curing in a humidity chamber (98% RH) for 2 days and a further 26 days in a normal laboratory environment.
- Curing in the humidity chamber during 28 days.

The concrete samples were cylindrical in form, of 75 mm diameter (0.0044 m^2 area) and 50 mm height.

In addition to composition, w/c ratio and curing process, water content in pores has a significant influence on gas transport. Therefore, three relative humidity ratios were studied: approximately 50% (laboratory environment), 75% and 83% RH. Samples were conditioned following the natural method specified in EN 83966 [8] for both 75 and 83%, replacing NaCl for KCl in this last case.

3.2 *Diffusion Coefficient*

The experiments to obtain diffusion coefficients were carried out using an airtight cell (diffusion cell, Fig. 1) in which the concrete sample was placed at the open end of the chamber sealing the joint between the sample and the enclosing cell. The chamber was ventilated with nitrogen until a null oxygen concentration was achieved and then the evolution of oxygen concentration within the chamber was monitored, which increases because of oxygen diffusion from the air through the sample.

Several airtightness tests were carried out. Firstly the airtightness of the whole cell is tested. In order to do this the chamber is closed, ventilated with nitrogen until a null oxygen concentration is achieved and then the evolution of oxygen concentration within the chamber is monitored. Results of this monitoring show that oxygen concentration within the chamber does not increase. Therefore it can be derived that the cell is airtight. Secondly the airtightness between the sample and the cell is tested and assessed. A diffusion test is performed with a capped sample (with double layer of aluminium tape). Results of the monitoring show that a small amount of oxygen is able to enter the cell. The air ventilation rate is calculated and compared to the limit value 3.78×10^{-3} h^{-1}established in ISO 11665-10 [9] draft. The calculated ventilation rate is smaller than the limit value, so the airtightness is considered adequate and the leak dismissed.

Diffusion coefficients D were obtained in stationary state using first Fick's law:

$$J = -D\frac{\partial n}{\partial x} \tag{3}$$

where:

J Oxygen diffusion flux [mol/m^2·s],
$\frac{\partial n}{\partial x}$ Oxygen concentration gradient [mol/m^4]

This expression, applied to our experimental case, results in the following form:

$$\frac{\left(N_{O_2}\right)_{t_{n+1}} - \left(N_{O_2}\right)_{t_n}}{S \cdot (t_{n+1} - t_n)} = -D\frac{\frac{(\overline{N}_{O_2})_{C_E}}{V_{C_E}} - \frac{(\overline{N}_{O_2})_{C_R}}{V_{C_R}}}{L} \tag{4}$$

where:

$\left(N_{O_2}\right)_{t_n}$ number of oxygen moles in reception chamber in time n, [moles],
S exposed surface of sample, [m^2],
$\left(\overline{N}_{O_2}\right)_{C_i}$ average number (in time interval n, n + 1) of oxygen moles in chamber i: E source y R reception, [moles], that is obtained experimentally,
V_{C_i} volume of chamber i, [m^3],
L height of sample, [m]

In our experimental case the test can be done using the constant source of oxygen concentration in air, so a source chamber is not needed and oxygen molarity in air can be taken as a constant (9.3 mol/m^3). Therefore Eq. (6) becomes:

$$\frac{\left(N_{O_2}\right)_{t_{n+1}} - \left(N_{O_2}\right)_{t_n}}{S \cdot (t_{n+1} - t_n)} = -D\frac{9.3 - \frac{(\overline{N}_{O_2})_{C_R}}{V_{C_R}}}{L} \tag{5}$$

4 Results

Results include experimental values of oxygen diffusion and permeability coefficients, as well as porosity of studied types of concrete and water contents.

4.1 Porosity and Related Parameters

Porosity has been determined (according to general procedure) with dry samples. The obtained average porosity values are represented in Fig. 2 for the four studied types of concrete. Types of concrete have been named as follows: cement amount (Kg/m^3)-W/C ratio-days of curing in humid chamber.

As expected, porosity is bigger for concrete with less cement, bigger w/c ratio and a shorter curing in the humid chamber. A better curing process (longer time in the humid chamber) reduces porosity, as well as a lower w/c ratio and a bigger amount of cement.

4.2 Permeability Coefficients

Average values for oxygen permeability coefficients are shown in the following table (Table 2).

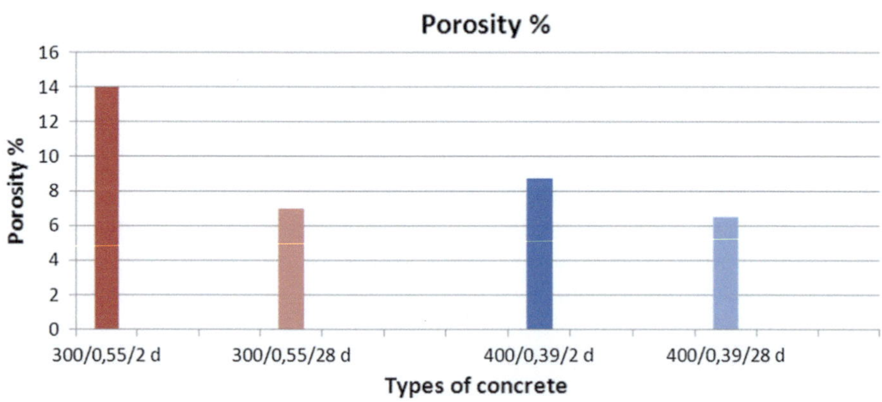

Fig. 2 Global porosity (%)

Table 2 Oxygen permeability coefficients (m^2)

	Relative humidity		
Concrete type (kg/m^3-W/C-curing)	50%	75%	84%
300-0.55-2 days	6.73E−20	2.77E−19	1.95E−19
300-0.55-28 days	2.35E−20	2.16E−20	1.10E−20
400-0.39-2 days	3.43E−20	9.02E−20	5.07E−20
400-0.39-28 days	2.19E−20	1.00E−20	7.61E−21

4.3 Diffusion Coefficients

The proposed methodology for measuring the gas diffusion coefficient provides values for the oxygen diffusion coefficients that are within the range of those proposed by the literature (10^{-7} a 10^{-8} m^2/s) and are consistent with the other studied parameters. Average values are included in the following (Table 3).

5 Discussion

The objective of the study is to provide a model for the calculation of the oxygen diffusion coefficient (D_{O_2}) from other concrete parameters of easier experimental determination. Those parameters have been identified as concrete porosity (ε) and permeability coefficient (K).

From the results of D_{O_2} and K for the three studied HR it can be observed that there is a qualitative correlation between them.

From the experimental values of K and D_{O_2}, a particular value of n or n' exponent in the Eqs. (1) and (2) was obtained for each type of concrete with its specific ε. With these particular n or n' values, regression analyses were performed to provide an expression relating n or n' to ε in a general way.

In addition to this, Eqs. (1) and (2) can be expressed in the following way:

$$D_{O_2} = e^{\frac{\ln(K)}{n'}} \tag{6}$$

$$D_{O_2} = e^{\frac{\ln\left(\frac{K}{a}\right)}{n}} \tag{7}$$

Therefore, with the expression of n or n', through inverse analyses it has been proposed a general expression that allows obtaining analytically D_{O_2} from experimental values of ε and K for Ordinary Portland Cement concrete in any condition.

Table 3 Oxygen diffusion coefficients (m^2/s)

	Relative humidity		
Concrete type (kg/m^3-W/C-curing)	50%	75%	84%
300-0.55-2 days	1.95E−07	1.70E−07	1.75E−07
300-0.55-28 days	4.85E−08	4.80E−08	1.33E−08
400-0.39-2 days	1.02E−07	1.05E−07	1.09E−07
400-0.39-28 days	4.05E−08	4.00E−08	1.00E−08

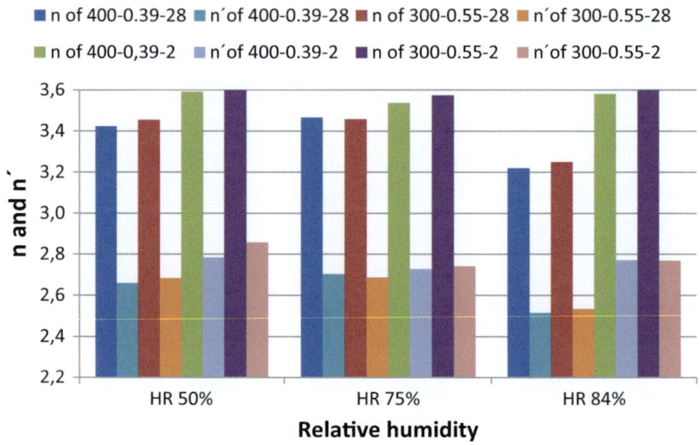

Fig. 3 n and n' values obtained from experimental values of D_{O_2} and K

5.1 *Particular Values of* n *or* n′ *Obtained from Experimental Values of Diffusion and Permeability Coefficients*

The n and n' values shown in Fig. 3 have been obtained for the different types of concrete (cement amount, w/c ratio and curing process) and humidity contents. As it has been mentioned above a is a constant which in our study case (with A $=$ 0.0044 m^2 and $D_{O_{2}o} = 2E - 05\,m^2/s$) takes the value of 439453.12.

5.2 *Expression Relating* n *and* n′ *to Porosity*

Several regression analyses have been performed to obtain expressions that allow obtaining n and n' from porosity (8 and 9).

Taking into account that 50% RH conditioning is the easiest to achieve because is done at laboratory conditions the analyses have been performed only for this water content. Figures 4 and 5 show the chosen regression expressions (8) and (9) for each case.

$$n = 0.35 \cdot \ln(\varepsilon) + 4.41 \quad R^2 = 0.94 \tag{8}$$

$$n' = 0.25 \cdot \ln(\varepsilon) + 3.37 \quad R^2 = 0.93 \tag{9}$$

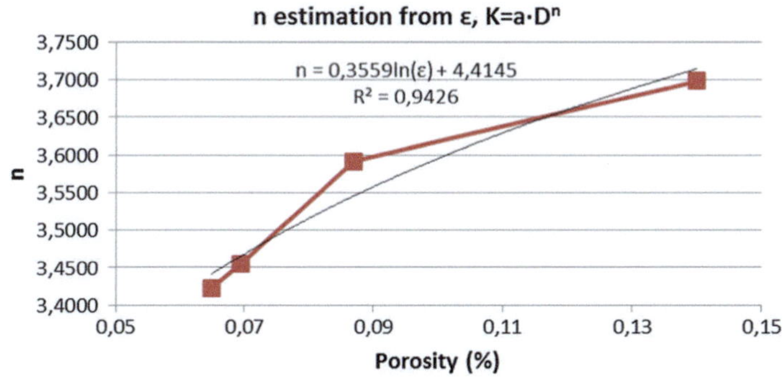

Fig. 4 Estimation of n from global porosity using $\mathrm{K} = \mathrm{a} \cdot D^{\mathrm{n}}_{O_2}$

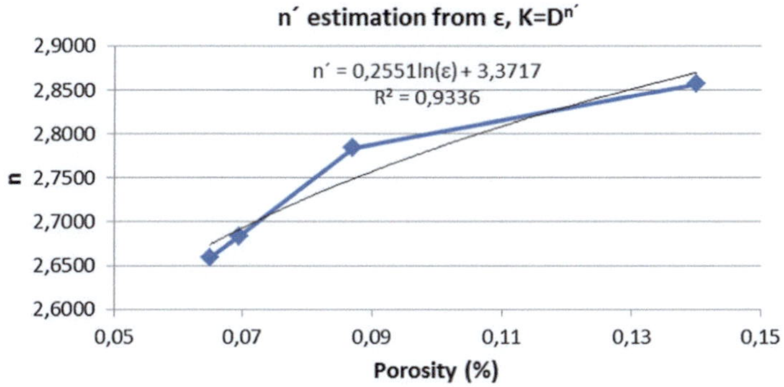

Fig. 5 Estimation of n' from global porosity using $\mathrm{K} = D^{\mathrm{n}'}_{O_2}$

5.3 General Expressions to Estimate Oxygen Diffusion Coefficient from Permeability and Porosity

Diffusion coefficient can be obtained from relations (6) and (7) with experimentally obtained K and using the estimated n and n' from the expressions (8) and (9). Expressions (6) and (7) turn respectively into:

$$D_{O_2} = \mathrm{e}^{\frac{\ln\left(\frac{\mathrm{K}}{\mathrm{a}}\right)}{0.35\cdot\ln(\varepsilon)+4.41}} \tag{10}$$

$$D_{O_2} = \mathrm{e}^{\frac{\ln(\mathrm{K})}{0.25\cdot\ln(\varepsilon)+3.37}} \tag{11}$$

These expressions enable diffusion coefficients to be directly estimated from the experimental permeability coefficients and porosity.

Figures 6 and 7 show the estimated diffusion coefficients and the experimental values.

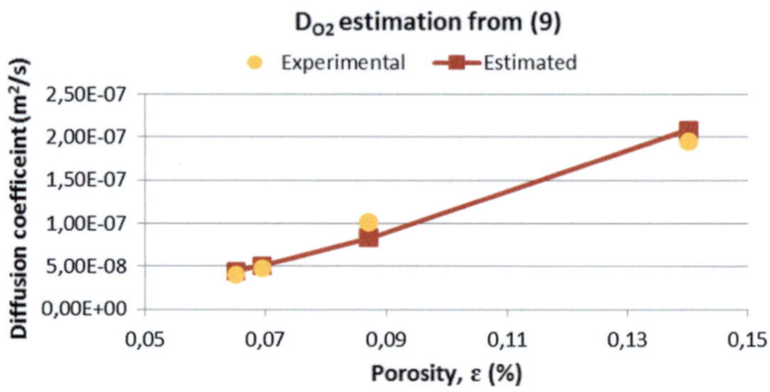

Fig. 6 Estimation of D_{O_2} from global porosity using $K = a \cdot D^n$

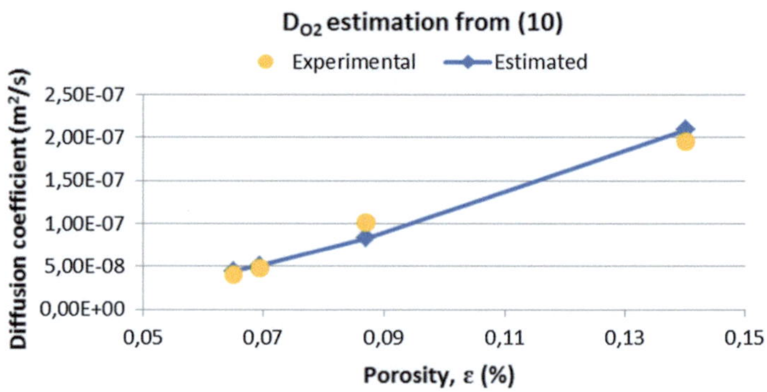

Fig. 7 Estimation of D_{O_2} from global porosity using $K = D^n$

6 Conclusions

The conclusions of the study are:

1. A gas diffusion cell has been designed and manufactured that, after several adjusts and tests, has proofed to be air tight.
2. An arrangement for measuring gas diffusion coefficient in concrete has been proposed that provides values for oxygen diffusion coefficients that are within the range of those proposed in the literature (10^{-7} a 10^{-8} m^2/s).
3. Different correlations between parameters related to gas transport in concrete were analysed. Several regression models were proposed to estimate oxygen diffusion coefficients from concrete porosity and permeability coefficient.
4. The use of these models will facilitate the estimation of oxygen diffusion coefficients from parameters that are easy to measure experimentally such as porosity according to ASTM-D4404-10 and permeability coefficient according to EN 83981: 2008.

5. These proposed models are exclusive for Ordinary Portland Cement, although they could be extended to other types of concrete, because they are based on porosity, which can be determined for any type of concrete.
6. The models are based on concrete porosity because it is easier to measure than tortuous porosity and their determination coefficients are similar to those obtained with tortuous porosity, even better in one case.
7. Two expressions are proposed to estimate oxygen diffusion coefficients in Ordinary Portland Cement concrete from permeability coefficient and global porosity ranging from 6.4 to 13.9%:

$$D_{O_2} = e^{\frac{\ln\left(\frac{K}{a}\right)}{0.35\cdot\ln(\varepsilon)+4.41}} \text{ or } D_{O_2} = e^{\frac{\ln(K}{0.25\cdot\ln(\varepsilon)+3.37}}$$

8. These models can be useful to characterize concrete lifespan as well as the performance of concrete as a radon barrier.
9. Further research must be performed to study the validity of these expressions in a different range of global porosity.

References

1. Gaber K (1988) Influence of mix proportions and components on the diffusion coefficient and the permeability of concrete. Darmstadt Concrete 3:39–48
2. Gräf HY, Grube H (1986) Influence of the composition and the curing of concrete on the gas permeability. Betontechnische berichte. (In German only)
3. Gaber, KY, Schlattner, E (1997) Final report of a research project Investigation of the pore structure of concretes with low radon permeability and possibilities for improving building materials with a high radon permeability. Sächsisches Staatsministerium für Wissenschaft und Kunst. (In German only)
4. Klink T et al (1999) Characterisation of the gas transport properties of porous materials by determining the radon diffusion coefficient. Mater Struct 32:749–754
5. European Committee for Standardization. Concrete durability. Tests methods. Determination to gas permeability of hardened concrete. EN 83981: 2008
6. ASTM-D4404-10 Standard Test Method for Determination of Pore Volume and Pore Volume Distribution of Soil and Rock by Mercury Intrusion Porosimetry
7. Andrade C et al (2012, April) Calculation of tortuosity factor for the model based in concrete resistivity. In: Second international conference on microstructural-related durability of cementitious composites, 11–13 April 2012. Amsterdam, The Netherlands
8. European Committee for Standardization (2008) Concrete durability—test methods—conditioning of concrete test pieces for the purpose of gas permeability and capillary suction tests. EN 83966
9. International Organization for Standardization Air: radon-222-Part 10: determination of diffusion coefficient in waterproof materials using activity concentration measurement. ISO/CD 11665-10. ISO, 2013

Petrographic Study of Siliceous Aggregates. Parametric Calculus of the Reactivity with the Alkalis

R. García-Rovés Loza, E. Menéndez Méndez and N. Prendes Rubiera

Abstract The microstructural stability of some components of aggregates is one of the determinants of the durability on concrete structures, regardless of their characteristics and intrinsic properties. Some pathologies such as alkali-aggregate reaction are linked with certain siliceous phases. Depending on its reticulated structure, the siliceous phases can provide siloxane groups to the media and promote nucleation of expansive gels, favouring the deterioration of concrete. The petrographic identification of the stress state is based on the physical property of the light polarization, which give defined optical results in deformed crystals. The digital imaging techniques allow to extract and to quantify these areas, establishing quantitative parameters that correlate exhibition area of siliceous aggregates edges with areas of fixation and development of expansive products.

1 Introduction

The aggregates are the main components in concretes and mortars with percentages ranging between 60% [1] and 80% [2]. The instability of the aggregates is analysed with a series of tests mainly petrographics [3]. These tests are aimed to identify and characterize components and reagents identify potential mechanisms associated with the new conditions of the interstitial solution (molarity, pH variations, relative humidity, water cement reaction, etc.) that would affect the durability of concrete and essential mechanical properties for the performance required.

Therefore the definition of aggregate as granular construction materials was completed with petrological and mineralogical considerations. In the definition was included cationic migration and fixation criteria and potentially harmful phases for

R. G.-R. Loza (✉) · E. M. Méndez
Institute of Construction Sciences "Eduardo Torroja"—IETcc, Madrid, Spain
e-mail: ricardo@ietcc.csic.es

N. P. Rubiera
Centre of Studies and Experimentation of Public Works (CEDEX) (Ministry of Public Works, Spain), Madrid, Spain

C. Andrade et al. (eds.), *Service Life and Durability of Reinforced Concrete Structures*, RILEM Bookseries 17, https://doi.org/10.1007/978-3-319-90236-4_8

stability and correct dosing of mixtures [4]. Although these criteria and phases not involved in setting and hardening reactions, themselves are the final determinants of technological properties such as mechanical strength, creep and abrasion [5].

Studies on the behavior of some aggregates for some specific physicochemical conditions of concrete have demonstrated their reactivity to alkalis, generating highly expansionary phases that can sometimes lead to the collapse of structures and a significant deterioration [6, 7].

Of the different reactions associated with expansive processes and their leachables products, such as generating of secondary ettringites [8], and brucite [9], the alkali-aggregate reaction (AAR) is the most aggressive and complex, including its two variants (alkali-silica (RSA), or alkali carbonates (RCA). These reactions are caused by interaction mechanisms and crystallochemical mismatches favored by stress states and deformation grades which present the aggregates.

2 Petrography of Quartz. Theoretical Principles

The reactions more complex in concrete are those involving the mineralogical phase of silica in its various forms and states, so much for your specific casuistry as its reactive behaviour.

2.1 Characteristics and Properties

The most common mineral in nature is quartz and, as such, is the phase most abundant in the majority of the aggregates, both fine and coarse. Quartz aggregate is used as materials in concrete mixtures.

Crystallographically, quartz consists of tetrahedra SiO_2 (silica), although its lattice structure varies from one species to another depending on the geochemistry of formation. The crystal lattice, at low temperature (<573 °C) is trigonal, whereas when reaching (<867 °C) becomes hexagonal. The Quartz polymorphic phases, such as cristobalite (in form αC and βC are tetragonal and cubic, respectively) and tridymite (hexagonal) appear at high temperatures (>1600 °C), as chalcedony (trigonal), opal, silica colloidal hydrated (triclinic) and flint (Fig. 1).

Reticular variability implies conceptually which the optical indicatrix is different in one case or the other by symmetry. The optical indicatrix is a virtual figure of mathematical formulation that draws the path of light passing through a crystal. Thus, low temperature quartz, seen by polarized microscopy, give a specific and light response different to high temperature quartz.

Consequently, changes in the crystal lattice (deformation, dislocation, packing defects, strains or tears), will affect the optical indicatrix and thus, to the microscopic image.

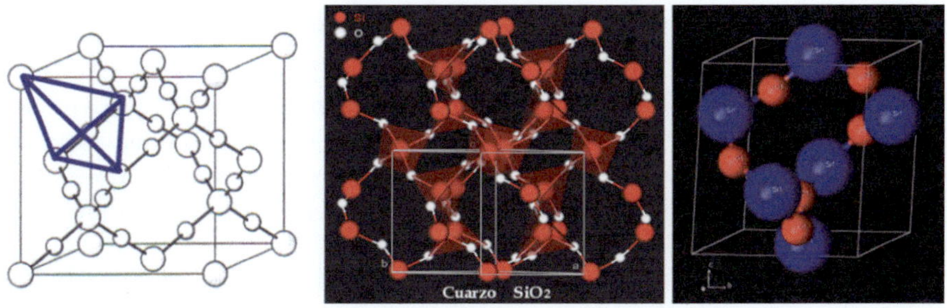

Fig. 1 Reticular structure of quartz polymorphs (Q), high and low temp, and tetrahedron base of glass (SiO_2) in chains (tectosilicate)

Fig. 2 Micrographs with polarized light of a thin section. Quartz with straight extinction in a matrix of mica and feldspar (left). Quartz with ondulatory extinction in a matrix of paste, quartz, mica and feldspar (right)

2.2 *Optical Identification of Quartz (Q)*

Petrographically the quartz gives a very specific behavior and it has its interference colors and textures particulars when is seen by polarized microscopy (Fig. 2).

The symmetry group and crystal habit are associated with the mineral phase reactivity due to the ability to fix and incorporate foreign elements into its structure. Thus a quartz with ordered lattice (trigonal or hexagonal), have a low capacity for cationic fixation (non-reactive). Quartz with lattice highly disordered (very low symmetry, triclinic amorphous type such as opals or chalcedony), have a greater susceptibility to produce those reactions of crystallochemical settings (high reactivity).

This petrological argument explains the silica behavior (SiO_2) and the proposed petrological classification of each mineralogical phase reactive quartz [10] based solely on statistical criteria and modal frequency distribution of the types of quartz with respect to its lattice state.

3 Potential Reactivity Mechanisms in Concretes

The siliceous aggregates with open structure of tetrahedra with Si–O–Si free groups and in a pH ambient very basic (pH $\approx$ 12), within in a interstitial solution rich in solubles ionic species (Na^+, K^+ y Ca^{2+}), react to produce silanol structures (Si–O–Si + $H_2O \rightarrow Si^{/OH\cdots OH\backslash}Si$). According to the Lewis definition, these structures are acidified and are added to pore solution, with the active process.

The silica have coordination 4 and a position is already fixed by OH [Si–OH]. The silica is already capable of combining with the hydroxides, according to the stoichiometric ratio Si–OH + $OH^- \rightarrow$ Si–O + H_2O, releasing water and leaving uncompensated species Si–O–, negatively charged. In this ambient, the rest of positive cations of the pore solution shall be fixed to compensate the charges balance, according with the formulation of [11] which utilize the Na_2O:

$$H_{0.38}SiO_{2.19} + 0.38\ Na_2O \leftrightarrow Na_{0.48}SiO_{42.19} + 0.38\ H_2O$$

From this point (Poole 1992), the process is repetitive, affecting to the rest of the siloxane bridges that pass into solution, according to Si–O–Si + $2OH^- \rightarrow$ Si–O^- + $^-$O–Si + H_2O. When this process crosses a threshold (pessimum effect), the silica is disintegrated or solubilizes in the form $H_2\ SiO_4^{2-}$, forming hydrophilic solid gel (Fig. 3), of variable composition (type silicone calc-alkaline), with immediate effect fissure as multiple gratings. In microscopy, the gels appear cracked when the sample is dry, evidence of structural water avidity.

3.1 *Petrographic Mechanisms*

Petrographic investigations with different species of SiO_2, which linking reactivity of quartz with reticular state, were addressed by [12, 13]. These authors conducted the research by light microscopy and from a qualitative point of view.

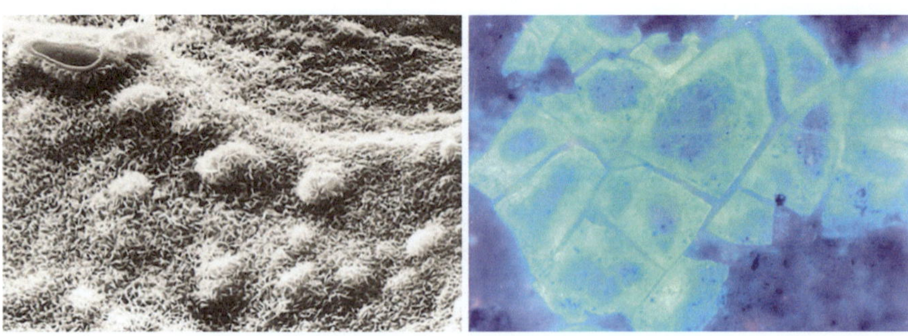

Fig. 3 Aspect of expansive gels with nucleation sites around the aggregates by SEM (left) and MOF (right)

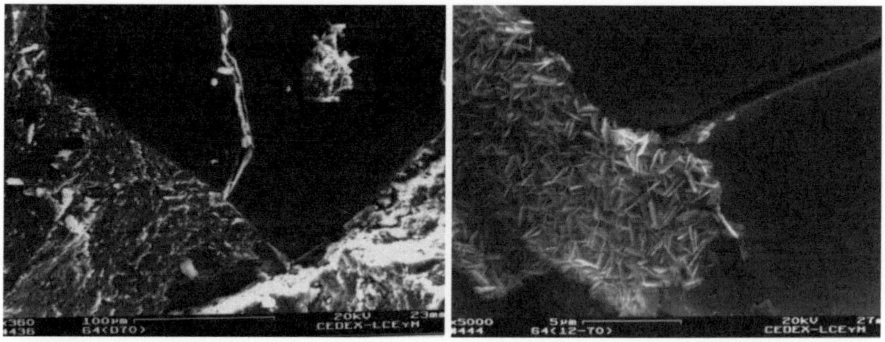

Fig. 4 Crack in quartz (SEM). It's filled with expansive products (gels), nucleated by the fixation of the hydroxyl groups which fixed cations of type alkaline and calc-alkaline

The measures were based solely in the undulatory extinction angles of quartz without addressing the problem (quantitative) from the boundary surfaces, or inner edges, which are the elements that would provide nucleation reaction.

The use of scanning electron microscopy (Fig. 4) in the analysis of the fracture surfaces confirms this phenomenology and also identifies the type, morphology and structure of the precipitate gel. From here it is possible to calculate the potential degradation of concrete.

For to establish criteria for monitoring and contrast to the works already mentioned, has resorted to optical polarization microscopy, using as criteria, parameters border.

4 Parametric Calculus of the Reactivity in Quartz

With the petrographic analysis will be defined the typology and stress state of each grain of quartz in the representative sample.

The petrographic microscopy displayed of each crystal its optical response. Through this process is discovered its lattice structure and thus its stress state, defining areas of the reticular network with the same crystallographic orientation.

4.1 Methodology

This method involves taking a images sequence with polarized light and rotating the microscopic platen (Fig. 5) of each grain of quartz (SiO_2). Through digital processing techniques are extracted morphological parameters or shape descriptors (area, perimeter, etc.).

Fig. 5 Images sequence in a quartz with varying degrees of rotation of the platen and areas with varying extinctions

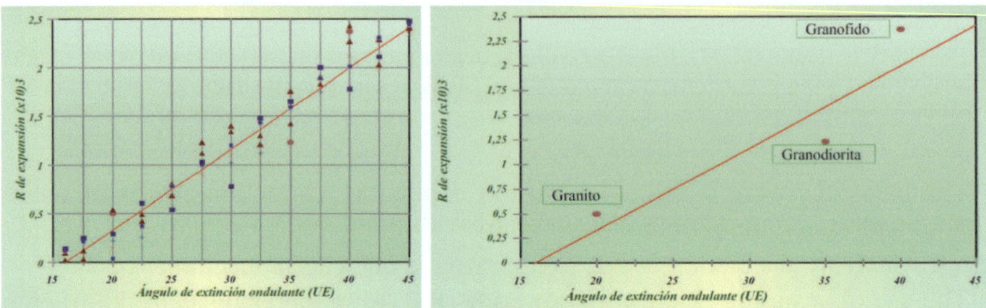

Fig. 6 Images sequence in a quartz with varying degrees of rotation of the platen and areas with varying extinctions

The first problem is to decide the minimum sequence of pictures to take, ranging from three [12] or [15] and four [16], at intervals ranging between 45° to 30° and whose purpose is to draw the areas of optical orientation to verify the phenomenon of undulatory extinction (UE). Some authors [14] correlated the petrographic type of different rocks types with the extinction angle, classifying, their potential reactivity, content and variety of quartz (silica). These correlations are in line with the classification given by [10] for siliceous rocks. These data [17] were correlated in the test method for potential alkali reactivity of Aggregates with the mortar-bar method [18], confirming these extinction angles. These correlations are illustrated in the Fig. 6.

In each image, the quartz grains are segmented and integrated into the same workspace, and the edges of each zone are extracted (Fig. 7). The result are the edges of extinction common areas of extinction (internal divisions), and the external boundary grain. Undeformed quartz gives no zoning. A greater deformation is observed more edges of areas of deformation.

4.2 Extraction and Processing of Data. Formulation

The undeformed quartz crystals have no internal zoning. The deformed quartz crystals have deformation internal areas. The images in Fig. 2 are illustrative. With this

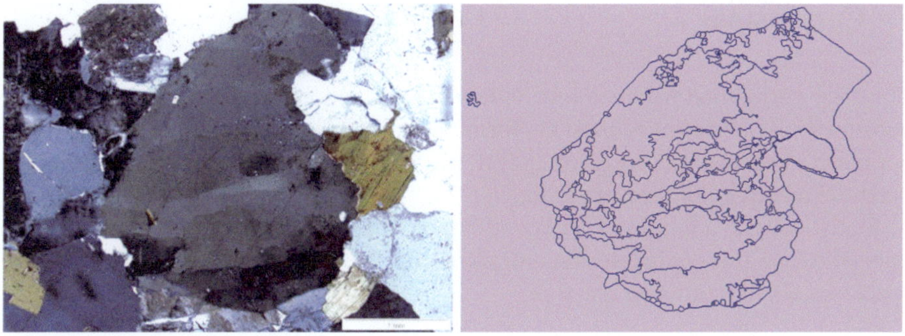

Fig. 7 Images sequence in a quartz with varying degrees of rotation of the platen and areas with varying extinctions

argument can establish a relationship between the grain boundary (perimeter or P_{ext}) and different internal subdivisions of contact (P_{int}) by the following expression (1):

$$I_{QZ} = \frac{P_{ext}}{(P_{\text{int}} + P_{ext})} \tag{1}$$

From this formulation, it can be stated that if the reactivity index of a quartz grain tends to $I_{QZ} \rightarrow 1$ are facing a stable crystal lattice, while if the reactivity index approximates to $I_{QZ} \rightarrow 0$ gives a distorted lattice structure (undulatory extinction); quartz with twins would have an index $0.75 < I_{QZ} < 0.65$. The study of a rock requires a statistical treatment which need developing a frequency histogram in which the ordinate represents the percentage and the abscissa the I_{QZ} value.

With this treatment, and from an order statistic, the aggregate can be considered potentially reactive if that is below the minimum required threshold calculated from the geometrical development of the diagonal, what for quartz is set in I_{QZ} <0,39. The reactivity index give a relationship spatial of the capacity siloxanes groups input to the pore solution. Is accepted which the phases are reactives from of the contact zones and structures able of to connect different points between constituents as pores or cracks. This index mark the relation between contact edges.

Each edge forms a chemical lability zone, with fixing capacity of cations and to form mineral phases. Some of these phases are potentially expansive (Figs. 4 and 5). The problem reduces to determine, quantitatively, these boundaries (perimeter of exposure) and that, once extracted, are compared with the expression established empirically from calibration curves testing of potential reactivity and relation surface/areas.

Recalculating the data of [12–14] and other authors, the resulting threshold between the quartz potentially reactive and the stable quartz is the following: I_{QZ} <0.38 [14] and for the case of [15], would reach I_{QZ} <0.25.

5 Conclusions

The clear correlation between low undulatory extinction angles and a crystal lattice deformed in a quartz grain, allows to formulate a spatial index (IQZ) between external perimeters and internal common perimeters of the deformation, that classified the potential reactivity of a quartz grain.

Statistically, if a family of quartz within of siliceous aggregates, is below a given threshold IQZ, this aggregate contain quartz capable of producing gels and expansive processes alkali-aggregate reaction.

References

1. Smith MR, Collis L (1994) Áridos. In: Áridos Naturales y de Machaqueo para la Construcción, 1st edn. Madrid, Colegio de Geólogos, p 435
2. Fernández Cánovas M (1996) Hormigón, 4th edn. Colegio de Ingenieros de Caminos, Madrid(Madrid)
3. ASTM C294. Standard descriptive nomenclature for constituents of concrete aggregates. ASTM Standards. Pennsylvania, ASTM Int
4. Henry R (1985) Cement and concrete terminology. Reported ACI. Detroit, Committee 116, p 58
5. O'Reilly VA (1997) Métodos para Dosificar Mezclas de Hormigón, p 111
6. Mather K (1966) Hardened concrete. Petrographic examination. s.l. ASTM 169A:125–143
7. Sims I, Hunt B, Miglio B (1990) Quantifying microscopical examinations of concrete for alkali aggregate reactions (AAR) and other durability aspect, s.l.: ACI, pp 267–287
8. Divet L, Ramdriambololona R (1998) Delayed ettringite formation: the effect of temperature and basicity on the interaction of sulphate and C–S–H phase. Cem Concr Res 28(3):357–363
9. Milanesi CA, Marfil SA, Batic OR, Maiza PJ (1996) The alkali-carbonate reaction and its reaction products an experience with argentinean dolomite rocks. Cem Concr Res 12–45
10. Menéndez E (2010) Análisis del hormigon en estructuras afectadas por reacción Álcali-Árido, ataque por sulfatos y ciclos Hielo-deshielo, pp 61–64
11. Glasser F (1992) Chemistry of the alkali-aggregate reaction. In: Swamy (ed) The alkali-silica reaction in concrete. Glasgow, s.n., pp 2–28
12. Dolar-Mantuani L (1981) Undulatory extinction in quartz used for identifying potentially alkali-reactive rocks. Cape Town, np, p 11
13. Sims I, Hunt B, Miglio B (1990) Quantifying microscopical examinations of concrete for alkali aggregate reactions (AAR) and other durability aspect, s.l.: ACI, pp 267–287
14. Grattan-Bellew PE (1993) Is high undulatory extinction in quartz indicative of alkali-expansivity of granitic aggregates?
15. Poole A (1992) Introduction to alkali-aggregate reaction in concrete. In: Swamy (ed) The alkali-silica reaction in concrete, pp 2–28
16. Martínez-Nistal A (1993) Discriminación, Cuantificación y Cartografiado de Componentes Petrográficos mediante Proceso Digital de Multi-Imágenes Microscópicas, Oviedo, pp 255: s.n
17. Swamy RN (1992) Testing for alkali-silica reaction. In: T. L. Ltd (ed) The alkali-silica reaction in concrete, 1st edn. Glasgow, Blackie and Son, pp 54–95
18. Dehills SM, Corvalan J (1964) Undulatory extinction in quartz grains of some chilean granitic rocks of differents ages. Geol Soc Am Bull (75):363–366

Chloride-Induced Depassivation of Steel in Concrete—Influence of Electrochemical Potential and Anodic Polarization Level

Marc Kosalla and Michael Raupach

Abstract Exploring the depassivation process of reinforcing steel in concrete and the chloride threshold level (CTL) is a major research challenge. The depassivation process is influenced by a wide variety of parameters. One of these is the steel potential, more precisely its spatial distribution, its variations and the resulting differences and polarization states. For certain electrochemical test methods sometimes an external polarization is applied, but the influence on naturally ongoing processes like the depassivation of the steel has remained largely unexplored. When determining the time of depassivation of reinforcing steel, in conjunction with determining the CTL, the application of an external anodic polarization is an established test setup. At the Institute of Building Materials Research (ibac) of the RWTH Aachen University the effect of external anodic polarization to different levels on the time to depassivation and on the CTL is explored. The current investigations show that there is no clear effect on the depassivation behavior for anodic polarizations up to +550 mV versus free rest potential in the passive state.

1 Introduction

Worldwide in technical guidelines for reinforced concrete (RC) structures the chloride concentration in the concrete adjacent to the steel rebar surface is considered to be the only, or at least the most decisive parameter for chloride-induced corrosions issues. Nowadays, however, it is well-known that the depassivation process of reinforcing steel in concrete is influenced by a wide variety of parameters and that the chloride concentration is just one of these. For summarizing the most important influencing factors it can be reduced to the characteristics of the steel/concrete interface (SCI) and the environmental conditions, i.e. the material parameters and resulting characteristics of the ambient concrete. However, for specifying this in greater detail, it is difficult to differentiate between direct and indirect effects. This is due to a large

M. Kosalla (✉) · M. Raupach
RWTH Aachen University, Aachen, Germany
e-mail: Kosalla@ibac.rwth-aachen.de

C. Andrade et al. (eds.), *Service Life and Durability of Reinforced Concrete Structures*, RILEM Bookseries 17, https://doi.org/10.1007/978-3-319-90236-4_9

number of influences affecting each other. Some parameters are strongly correlated, and consequently it is not really clear which parameter characteristics actually cause depassivation and which are just minor contributing side effects.

A possible influencing factor which is subject to some discussion is the steel potential respectively the existing potential differences sometimes in conjunction with an external polarization, for instance when using laboratory specimens. However, like the depassivation process, the potential itself is also subject to several other influencing factors. Due to the large number of parameters the mutual interferences are mostly difficult to evaluate. Thus finally the question remains, whether the free rest potential distribution arising in practice has a direct influence on depassivation occurrence or whether potential differences between certain areas on the steel surface are just another side effect of varying boundary conditions (caused by concrete and SCI characteristics), which are themselves decisively affecting the depassivation process. Furthermore, it is questionable whether an external polarization will affect the depassivation process and whether an external, potentiostatic polarization is comparable to natural polarization states.

2 Objectives and Background of the Investigations

The role of the electrochemical potential regarding the depassivation process of steel embedded in concrete is complex and thus difficult to define. On the one hand, in RC structures spatial variations in potential levels are an essential requirement for generating anodic and cathodic areas and therefore for initiating and sustaining an active corrosion process. Also in the passive state, under certain conditions, significant potential differences may be present in a concrete structure due to the inherent large extent of spatial heterogeneities [1]. It is obvious that in some way the electrochemical potential must be involved in the depassivation process, but until now the specific nature of the impact on the time and location of the first depassivation spot in a RC structure has not been clearly demonstrated. On the other hand, it is also obvious that the spatial potential distribution is just the result of the characteristics of the steel surface, the SCI and the ambient concrete [1]. The parameters involved mostly underlie mutual interferences. This leads to the question, whether certain conditions of the SCI and the ambient local environment induce a depassivation process, or if these conditions lead to certain potential features causing the depassivation.

The work presented in this paper was carried out in the framework of a research project funded by the German Research Foundation (DFG). The project aims to investigate the chloride threshold level (CTL) for different binder combinations and at quantifying essential influencing factors on the CTL on a macro scale. Even though there are many influencing factors for depassivation issues, the CTL remains an important parameter for the assessment of the probability of depassivation occurrence and for durability considerations in service life calculations. In this paper, the study of the effect of a varying external polarization level to cubic laboratory specimens on the CTL is presented. In the long term, this is one step towards the development of a suitable test method for determining the CTL, which is strongly needed as an approach of worldwide research work.

In another test series performed in parallel, potential distributions and polarization states have been investigated for specimens with an entirely passive surface of the reinforcing cage and dimensions representative for actual practice. The highest determined potential difference was found to be 150 mV for concrete made with ordinary Portland cement (OPC) whereas a maximum potential difference of 650 mV was detected for concrete made with blast-furnace slag cement (BFSC). The highest measured depolarization was only 17 mV for OPC and 352 mV for BFSC [1]. In this context, depolarization means the potential shift occurring when de-coupling steel segments exhibiting different potentials. These outcomes are considered as indicative values for creating suitable measurement settings for further investigations of the CTL with specimens under potentiostatic control (PC) in order to simulate electrochemical conditions, which are relevant for situations occurring in practice.

3 Theoretical Background

3.1 The Conceptual Approach of CTL

The concept of CTL or critical chloride content assumes that steel embedded in uncarbonated concrete is protected from corrosion as long as a certain chloride concentration in the concrete adjacent to the steel is not exceeded. In Fig. 1 the CTL is assumed to be 0.5 wt%/cement and in the situation depicted this level is not achieved at the depth of the reinforcement (25 mm). The actual chloride profile exceeds the threshold level only in the near concrete surface region up to a depth of approximately 17 mm. Thus, the steel having a higher cover depth will be in a passive condition here. In theory, when the actual chloride concentration at the depth of the reinforcement exceeds the CTL, depassivation will occur.

Worldwide technical guidelines usually define a certain threshold level, mostly between 0.4 and 0.6 wt% with respect to the binder mass (wt%/cem) [2, 3], which is implicitly assumed to be independent of other parameters and boundary conditions, e.g. the binder type and the moisture content. In the course of time, this approach turned out to be too simplistic, because many investigations made on the CTL in the last decades indicated that the corrosion-inducing chloride concentration is subject to a very wide variation due to a multitude of other influencing factors on the depassivation process. Amongst others, Breit [4], Angst [5] and Glass [6] provided overviews about the investigations made during the last decades.

The wide variability found is in turn attributed to several causes. On the one hand, material inhomogeneities and varying boundary conditions have a significant impact. Especially, concrete demonstrates wide variations in material properties leading to strong variations in the electrochemical characteristics. The binder type is already decisive for the pH-value of the pore solution, the pore structure and consequently

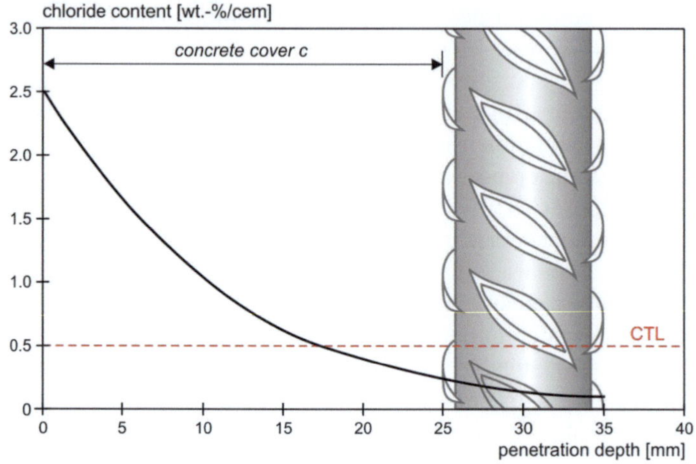

Fig. 1 CTL not reached here (if the intersection of cl-profile and dashed line is not in the concrete cover area, the steel is active in theory)

for the chloride transport and binding properties of the concrete. The w/b-ratio and the general binder content also affect the porosity and its structure. Furthermore, the natural inhomogeneity of a specific concrete composition in general can lead to high local performance deviations in a structure [1]. This is generally depending on the spatial distribution of aggregates, pores and defects. In particular inhomogeneities at the SCI are likely to have a considerable impact on the depassivation behavior [7]. In contrast, steel is generally a much more homogeneous material, but its post-treatment and storage conditions can also cause inhomogeneities or defects on the surface leading to non-uniform electrochemical characteristics and consequently to deviations regarding the CTL [8]. In addition, various types of post-treatment cause different microstructural characteristics of the steel surface [9].

On the other hand, various researchers designed and established partially very different measurement approaches, test setups and procedures to determine the CTL of laboratory specimens. These approaches include varying anodic and cathodic areas, cathodic materials, failure criteria, chloride ingress procedures and many other variations. Additionally, varying sampling and analyzing methods can cause further deviations, also regarding samples obtained from real RC structures in practice.

Thus, most CTLs determined in investigations made worldwide are actually not comparable. For a better assessment of the outcomes of determining the CTL, there is a strong requirement for investigating the quantitative effects of the main influencing factors. Moreover, a suitable standard test method would be urgently needed for achieving a better understanding of the depassivation process. The test procedure has to be relatively rapid, reproducible and practicable to find application and recognition in research as well as in practice.

3.2 *Influence of Electrochemical Potential and Polarization State on the CTL*

There are different measuring approaches regarding the determination of the time of depassivation varying in several categories, starting with the parameters affecting the failure criterion adopted and monitored for indicating depassivation. Besides following just the open-circuit potential (OCP) e.g. according to Angst [10] and Garcia [11], also the development of corrosion currents or the polarization resistances can be monitored. For a specific approach involving a nobler counter electrode (acting as cathode) an external anodic polarization (i.e. an externally applied d.c. voltage) is imposed in order to ensure a well-defined and reproducible electrochemical system [potentiostatic control (PC)] and the element current flowing from the rebar to the external counter electrode is monitored over time. This method was already applied by Breit [4]. However, the effect of this external interaction in the free electrochemical system on the depassivation process is not completely clear.

Dauberschmidt compared a number of investigations performed by two different researchers at the Institute of Building Materials Research at RWTH Aachen University (Fig. 2) [12] and noted a significant difference between the results obtained by Breit working with a high anodic polarization level up to +550 mV versus the initially determined steady-state rest potential [4] and results obtained by Harnisch just monitoring the OCP and the element current without any external polarization [7]. Based on the significant differences in the results Dauberschmidt [12] concluded, that there is an effect of external anodic polarization on the depassivation process decreasing the CTL. Breit [4] himself already perceived the same effect of external polarization for concrete with OPC, however he observed an opposite trend for BFSC.

Other researchers, e.g. Alonso [13], obtained contradictory results for externally polarized laboratory specimens, indicating no noticeable effect of the anodic polarization level, but just an influence of a cathodic polarization which is confirmed e.g. by Sanchez [14] and many others. Alonso also provides an overview of the polarization impact for several investigations [13] however without coming to any clear conclusions. Thus, further investigations to check the effect of an external anodic polarization to different levels are urgently needed. By checking the suitability of the procedure for a range of anodic polarization levels, the outcomes for different levels can directly be examined and compared for evaluating and maybe quantifying the actual influence.

4 Investigations

4.1 *Objectives of the Investigations*

In this paper, investigations on the effect of an externally induced anodic polarization and its extent are described. The investigations are aimed at examining the external

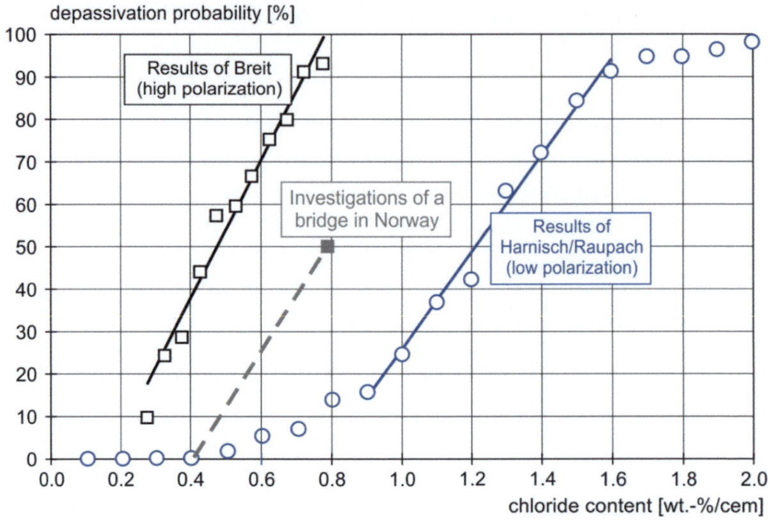

Fig. 2 Illustration of results from different measuring approaches according to [12]

polarization method regarding its suitability for determining the time to depassivation and quantifying the critical chloride concentration. The effect of various polarization levels is investigated, with the long-term aim of finding suitable test conditions for CTL investigations.

4.2 *Specimens and Materials*

For this investigation, 36 cubic specimens having a 150 mm edge length and a centrally embedded ribbed rebar have been produced (Table 1). The rebars have a diameter of 10 mm and are made of cold-rolled steel (B500A). For 18 cubes ordinary Portland cement (OPC, CEM I) has been used, whereas for the other 18 specimens blast-furnace slag cement (BFSC, CEM III/A) was used as the binder. The concrete mixes are designed to fulfill the requirements of environmental class XD3 according to EN 206 [3] (chloride-loaded exposure to wetting and drying). The cement content amounts to 320 kg/m^3 and the w/b-ratio is 0.45. In this test series, no superplasticizer has been used. But additionally, three specimens with OPC and a lightly deviating grain-size distribution have been produced with superplasticizer.

For casting the specimens, the rebars have been vertically fixed in wooden moulds (Fig. 3) as to achieve a high-quality SCI without any settlement or bleeding effects [15]. Demoulding of the specimens occurred one day after casting. After demoulding, the electrical connections of the rebars were prepared at the protruding end of the rebars which is not embedded in the concrete. Next an electrically isolating coverage was applied at both ends of the rods by Block Coat (from Co. Protector) for the

Table 1 Test schedule and polarization levels

Binder type	Polarization level versus rest potential (mv)	Number of specimens
OPC	+0	6*
	+100	3
	+200	3
	+300	3
	+400	3
	+550	3
BFSC	+0	3
	+100	3
	+200	3
	+300	3
	+400	3
	+550	3

*Three additional specimens produced with superplasticizer and modified aggregate size distribution

purpose of corrosion protection. After this treatment, the specimens were stored under water until an age of 28 d followed by storage in the climatic chamber at 20 °C and 95% relative humidity until an age of 90 d. After removing the specimens from storage, the concrete cover was reduced to 10 mm by sawing in order to reduce the time to depassivation to an acceptable duration. A synthetic tank was attached to the sawn concrete surface by means of polyurethane (Sikaflex) for allowing a permanent chloride solution exposure. Subsequently, the lateral specimen surfaces were sealed with epoxy resin as to prevent any influence of the environment regarding moisture content at these surfaces in order to simulate the conditions of a plane construction element. After preparing the specimens, a pre-wetting during 1–2 d was performed in order to avoid huge capillary suction effects and a possible immediate depassivation upon first chloride exposure.

4.3 Experimental Setup and Procedure

Before starting the electrochemical measurements, the upper concrete surface was exposed to a 3.0 wt% sodium chloride solution. A titanium mixed metal oxide mesh was placed on the concrete surface to serve as a counter electrode (C) and a manganese dioxide (MnO_2) electrode ERE 20 (Co. Force Technologies) is also placed in the solution acting as a reference (R) (Fig. 4). The titanium counter, the manganese reference and the steel rebar working electrode (W) were connected to one channel of the multichannel potentiostat, which has been developed at ibac for these experiments. The concrete specimens were placed on plastic spacers as a practical way to allow air ventilation from the lower side.

Fig. 3 Wooden mould with vertically fixed rebar

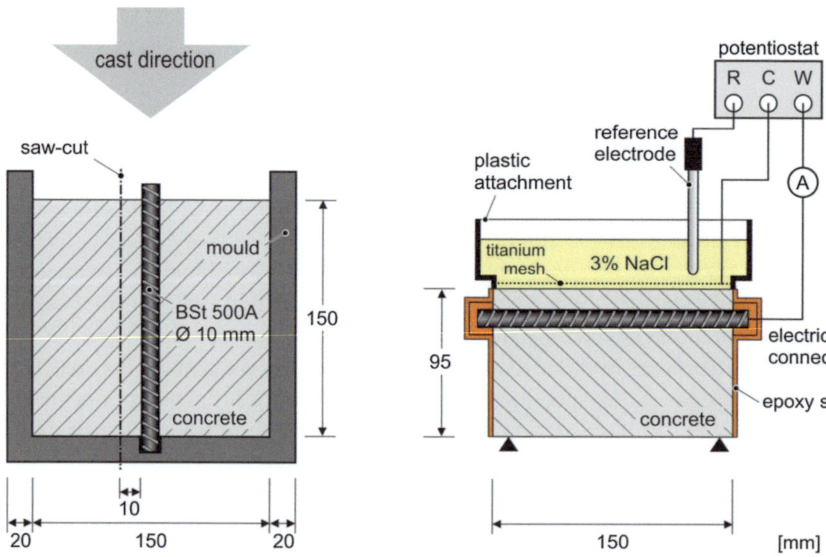

Fig. 4 Specimens and measuring setup

Initially, for every specimen the rest potential of the embedded rebar versus MnO_2 was determined over a time period of 2–3 weeks until an acceptable steady-state level (ΔU <3 mV/day) was achieved. After achievement of this rest potential, the rebars were subjected to potentiostatic control (PC). For both binder types, 18 specimens were subjected to six different potential levels, starting with potentiostatic control at the level of the rest potential. In the next level, the specimens are polarized to rest potential +100 mV. Then the anodic polarization level increases in steps of +100 mV. The highest polarization level was set to +550 mV (instead of +500 mV), because

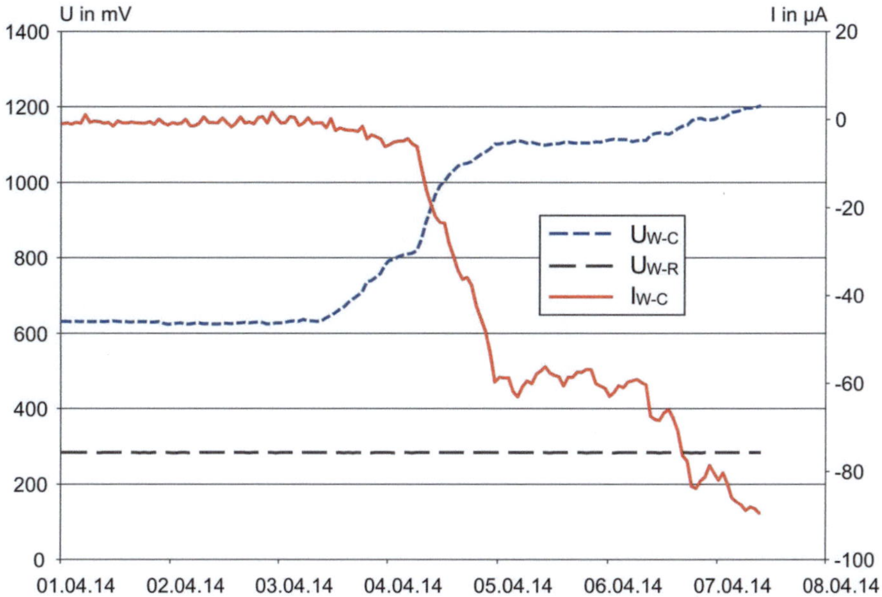

Fig. 5 Typical example of the development of measuring parameter in case of depassivation

this is the polarization level Breit employed [4]. The schedule for these investigations is given in Table 1.

During externally induced polarization the upper concrete surface was continually exposed to the chloride solution. The polarizing current I_{W-C} flowing as well as the potential difference U_{W-C} between the working electrode (W), thus the reinforcing steel, and the counter electrode (C) were continually monitored, while maintaining a constant anodic polarization level U_{W-R} versus the reference electrode (R). Due to the potentiostatic control the occurrence of depassivation can be detected by a significant increase of the element macro cell current I_{W-C} ($\leq 10\ \mu A$) accompanied simultaneously by an instantaneous potential shift ΔU_{W-C} [>100 mV in less than 4 h (for OPC)] between rebar (W) and counter electrode (C). As an example, Fig. 5 shows the development of these two electrochemical parameters, i.e. element current I_{W-C} and potential difference U_{W-C} between working and counter electrode in case of a depassivation event. The black dashed line illustrates the constant potential level U_{W-R} between working and reference electrode forced by PC.

As soon as a clear and sustained increase of the element current (depassivation criterion for this test arrangement: $I_{W-C} \leq 10\ \mu A$) was observed, the chloride solution was removed, all electrical connections were disconnected and the specimen was split along the rebar (Fig. 6). Thereupon the retrieved rebar was subjected to a visual observation to determine the location (depth and position) of the first corrosion spot and to characterize the concrete and SCI adjacent to the corrosion spot (Figs. 11

Fig. 6 Example of a specimen after splitting for visual observations

and 12). The "negative" current generated following depassivation is just due to sign convention of the measuring device.

4.4 Sampling and Determination of the Chloride Content

After splitting, smaller cubic concrete samples were removed from the specimen by dry-sawing. For every specimen, one cubic sample was extracted within the area of the first corrosion spot. For some specimens various locations were analyzed (compare Fig. 7). The removed cubic samples were bonded to stamps in order to fix them in a grinding machine (Fig. 8). This machine includes a controlled propulsion and presses the sample against a grinding wheel allowing concrete to be reduced to powder samples of 2 mm depth intervals. The CTL of the specimens was determined for a sample at the depth interval 9–11 mm. The resulting powder samples were analyzed by potentiometric titration according to [16].

5 Results

5.1 Time to Depassivation

The electrochemical measurements were finalized after 603 days of chloride exposure. After such a long duration of continuous exposure it was considered very questionable if depassivation of the steel rebar embedded in the passive concrete specimen was likely to occur in the near future. The depassivation times of the specimens which clearly indicated onset of corrosion are listed in Table 2. For the OPC

Fig. 7 Typical example for sample allocation after visual observation

Fig. 8 Grinding for obtaining samples for chloride analysis

specimens, for 12 out of 18 rebars, an active corrosion process was detected during these 603 d. For the BFSC specimens, for only one rebar out of 18, a depassivation process was indicated within this period.

5.2 *Chloride Analysis*

The determined corrosion-inducing chloride concentration of OPC specimens is in the range of 0.88–1.68 wt%/cem. Table 2 provides an overview of the determined CTL (chloride content at the upper rim of the rebar at the time of depassivation) for the depassivated specimens of the test series. Figure 9 illustrates these CTL values as a function of the imposed polarization state.

Table 2 Depassivation times, CTLs

Specimen	Polarisation versus rest potential (mv)	CTL (wt%/cem)	Time to depassivation
0-PZ-B-1	+550	1.04	175
0-PZ-B-2	+550	0.88	147
0-PZ-B-3	+550	0.94	109
0-PZ-400-1	+400	1.63	235
0-PZ-400-3	+300	1.60	373
0-PZ-300-1	+300	1.54	303
0-PZ-300-3	+300	1.37	541
0-PZ-200-1	+200	0.88	119
0-PZ-200-3	+200	1.22	512
0-PZ-100-1	+100	1.07	141
0-PZ-R-2	+0	1.68	384
0-PZ-R-3	+0	1.11	120
0-HOZ-200-1	+200	0.73	541

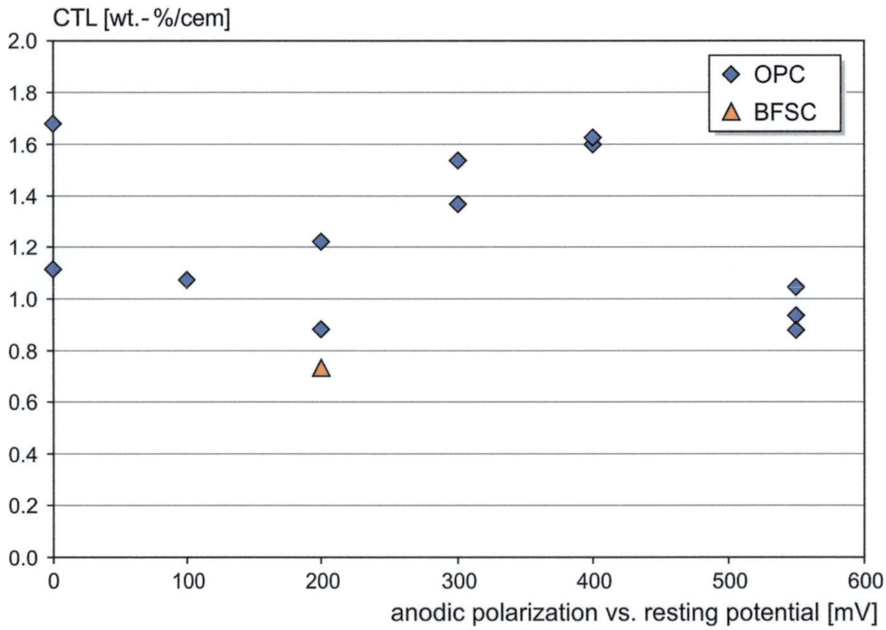

Fig. 9 Distribution of CTLs as a function of the anodic polarization level

For the non-corroded specimens, the measured chloride content at the reinforcement after 603 d of exposure is in the range of 1.11–1.69 wt%/cem for OPC, whereas for BFSC the chloride concentration ranges from 0.20 to 1.43 wt%/cem. These results are illustrated in Fig. 10. In five out of seven BFSC specimens the chloride concen-

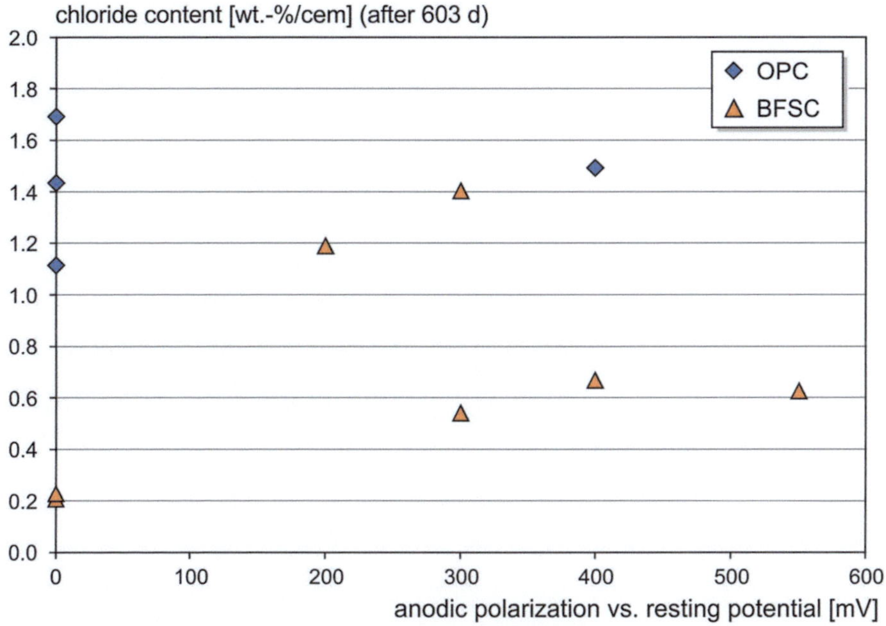

Fig. 10 Chloride concentration at passive rebars after 603 d of exposure as a function of the imposed anodic polarization level

tration exceeds 0.54 wt%/cem. It should be noted that the values reported for OPC include two of the specimens made with superplasticizer kept under PC at the level of the initially determined rest potential. In these specimens, the chloride concentration present at the reinforcement amounted to 1.44 and 1.11 wt%/cem, respectively.

5.3 Visual Observations

Visual observations were made in order to localize the first corrosion spot, to determine the actual cover depth at that spot and—if possible—to characterize the condition of the concrete as well as the condition of the SCI adjacent to the corrosion spot. The location is of essential importance for sampling of concrete material.

For all specimens demonstrating depassivation occurrence indicated by an element current increase (see Sect. 4.3), corrosion products are localized in the upper area of the rebar, at a depth interval of either 9–11 mm, 9–13 mm or 9–15 mm. In 3 out of 13 cases, no defects or abnormalities were detected adjacent to the corrosion spot (Fig. 11). In one case, a larger grain of aggregate was located next to the corrosion spot and possibly this had an impact on the depassivation process. In four cases, a larger pore is found next to the corroding spot and may be involved. At four further specimens, a grain and a pore are adjacent to the corrosion spot (Fig. 12). In another

Fig. 11 Corrosion initiation spot in the absence of defects (OPC, +300 mV)

Fig. 12 Corrosion initiation spot close to a pore (OPC, +300 mV)

specimen, no allocation was possible, because multiple spots having larger steel surface areas corroded.

Considering the specimens without any indication of depassivation (including the specimens made with superplasticizer), it has been observed that in 10 out of 26 specimens (2 out of 9 for OPC; 8 out of 17 for BFSC) corrosion products were detected at the SCI, although the electrochemical measurements did not indicate an active corrosion process. The amount of corrosion products is usually relatively small and comparable to single spots shortly after depassivation. The cover depth of these spots covers the whole range of the entire rebar diameter between 9 and 21 mm. Only in one specimen, the corrosion spot appears next to a pore, however in none of the other specimens defects or abnormalities at the SCI were detected.

6 Discussion

6.1 Evaluation of Chloride Contents

In general, the range of determined CTL values fits the expectations for a high-quality SCI. The CTL values are clearly higher than 0.5 wt%/cem which is the currently stated threshold value in German guidelines [2] and comparable to the findings of the investigations by Harnisch [7]. The CTL determined for the only depassivated BFSC specimen is found to be less than those found for all OPC specimens and higher than 0.5 wt%/cem, however it is difficult to evaluate a single result.

Comparing the different electrochemical conditions, no effect of polarization level can be observed—not even assumed. The deviations can be attributed to natural scattering phenomena due to the high inhomogeneities of the material (see 3.1). According to the findings of Kosalla [1], only the polarization levels "+0 mV" and "+100 mV" are considered to be of practical relevance for OPC structures, because higher polarization levels were not determined for OPC, in contrast to BFSC structures.

Considering the determined CTL values with respect to the general potential level versus SCE (standard calomel electrode) in the polarized state, neither a clear dependency can be observed. Due to the small range of initially determined rest potentials (−84 mV ± 17 mV vs. SCE) the illustration of results in Fig. 9 would also display nearly the same pattern regarding the distribution of results [with small deviations (shift by ±17 mV to the left or to the right)].

6.2 Evaluation of Visual Observations

Due to the intended uniform, high-quality SCI promoted by casting with vertical rebar orientation it was expected that corrosion starts at the upper rim, because the chloride concentration is usually higher the shorter the distance from the exposed concrete surface is. The observation of the rebar prints in the concrete confirmed this pre-assumption for all specimens with electrochemically indicated depassivation.

For the specimens which did not demonstrate an increase of the element current during the test duration, at several rebar prints small amounts of rust could be detected. At the steel surfaces, also corrosion products were detected, but only superficial corrosion without actual pitting corrosion spots was observed. The amount of corrosion products is similar to the amount at indicated depassivation spots. This leads to the assumption of the presence of a temporary active corrosion process. It is conspicuous that the majority of these corrosion spots is located in the border area of the rebar, where the concrete surface is actually covered by bonding material for the exposure tanks. Furthermore, the corrosion initiation spots are partially not at the upper rim of the rebar. However, these deviations do not provide an explanation why the corrosion activity was not indicated by increasing element currents.

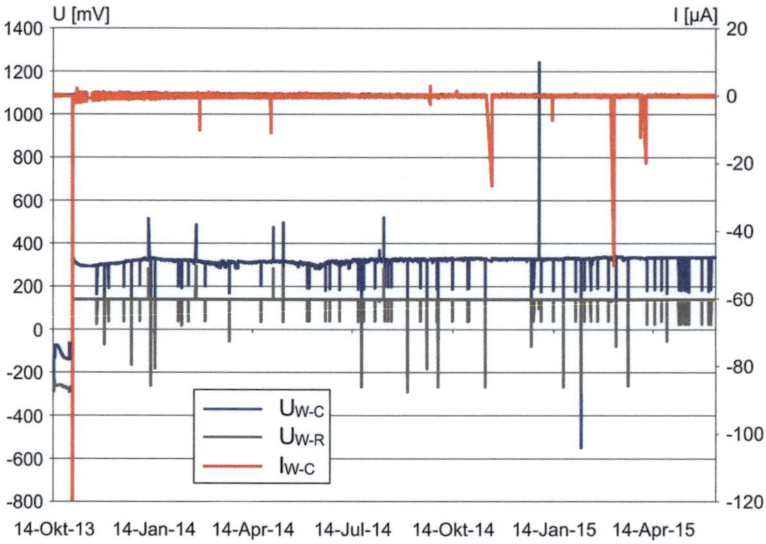

Fig. 13 Example of the development of measuring parameters for a specimen without indication of depassivation (BFSC; polarization: +400 mV)

6.3 *Evaluation of Electrochemical Measurements*

Considering the specimens without corrosion indication by a current increase, a number of possible explanations can be given for the presence of corrosion products. The most obvious explanation lies in the detailed examination of the element current development. By reconsidering the magnitude of the element current in the predefined passive current range, the failure criterion was rechecked. However, the measured currents were in the same range as those in passive specimens without any visual corrosion indication.

Figure 13 shows a typical development over time of the measured electrochemical parameters for one of these specimens. It should be noted that there are several short-time extreme value peaks of the element current I_{W-C}. These peaks resulted from short-term failures of the measurement system lasting up to 3–4 days. When restarting the measurements, in many cases the short-term response of the specimens was a direct and strong increase of the element current in many cases, usually reverting to the passive level within 30–60 min. The abrupt reapplication of the PC can possibly force severe electrochemical reactions of the system for reaching the applied conditions, leading to an active corrosion process induced by the impressed current. This is a common phenomenon which has been often observed at ibac when applying an external interference in an electrochemical system. Although these currents revert very quickly, such a reapplication of the PC can provoke a temporary rapid dissolution of iron resulting in the appearance of corrosion products.

Another explanation is the occurrence of a possible self-corrosion phenomenon, maybe also induced by the temporary breakdown of the measurement technique. Due to the resulting changes in the electrochemical system, the availability of the titanium mesh as cathode can be eliminated. Furthermore, by interrupting the externally induced polarization the rebar is allowed to create significant potential differences and corresponding cathodic areas along its surface. Thus, the development of self-corrosion activities on the steel surface is not unlikely. Another evidence of this phenomenon is the fact that the concerned rebars show corrosion activities in the border area, where the formation of cathodic areas is likely to be due to the absence (or at least due to a lower amount) of chlorides.

The explanation by a pre-corroded condition of the steel rebar before casting is not considered plausible, because the corrosion products are strongly adhering to the concrete after removal of the steel, more than to the steel surface. The determined chloride concentrations found at these spots are widely ranging from 0.39 to 1.50 wt%/cem. Because most of the affected concrete specimens were made with BFSC, it has to be considered whether the test procedure is generally suitable for investigations on steel corrosion in slag cement concrete.

6.4 Evaluation of Measurement Setup

It is obvious that a test duration of more than 1 year is not acceptable for a functional standard test method. Consequently, the pre-treatment of the concrete specimens as well as the boundary conditions have to be critically reconsidered. The measurement setup in general, including a rebar under PC with a low anodic polarization and an increase of the element current as failure criterion, was generally functional for OPC specimens. For BFSC specimens, however, the functionality of the method is questionable, because of the obviously corroding spots found in some specimens, although no sustained current increase was observed. However, it is likely that this may also be caused by other influences (see Sect. 6.3).

7 Conclusions and Outlook

7.1 Conclusions

In this study, only reinforced concrete with a high SCI quality—enabled by casting with vertically oriented rebars—is considered in order to avoid an impact of defects. The following conclusions can be drawn from this study:

- The corrosion-inducing chloride content (CTL) for depassivation of reinforcing steel in concrete, made with OPC according to XD3-requirements [3], was determined in the range 0.88–1.68 wt%/cem.

- A clear influence of an externally imposed anodic polarization on the depassivation behavior was not observed. Neither the CTL nor the time to depassivation seems to be affected by the polarization level.
- Considering the influence of the general potential level in the polarized state, neither a clear dependency was observed.
- For long-term studies with specimens under permanent PC, the stability of the measuring system is of essential significance in order to avoid confusing measuring data. Failure occurrence—causing an on/off-switchover regarding polarization—-may lead to sharp short-term reactions of the electrochemical parameters forcing temporary active corrosion currents and corresponding occurrence of corrosion products, independent of the presence of chlorides.

7.2 Outlook

Further investigations in the framework of this research project will be executed with a low level of anodic polarization (rest potential +200 mV). The applied experimental procedure has to be improved, since the test duration required to achieve depassivation is far too long and not suitable for a practical standard test method. Thus, the pre-treatment of the concrete specimens used should be modified in order to reduce the test duration. The failure criterion for indicating corrosion initiation may be reconsidered and adapted to the boundary condition (e.g. binder type and SCI quality). Thus, the response of current and potential level to depassivation events continues to be observed in detail in any specific case.

Thus, in future tests, the effect of varying drying duration after storage in the climatic chamber on the time to depassivation will be investigated for different concrete covers. The objective of these tests is to adjust and optimize the test duration for a prospective standardization of the test conditions. Additionally, the influence of specimen size, the further variation of binder combinations by addition of fly ash and silica fume, as well as the effect of a reduced SCI quality will be subject of future investigations made in the framework of this research project.

Acknowledgements The authors would like to thank the German Research Foundation (DFG) for the financial funding of the project.

References

1. Kosalla M, Raupach M (2016) Mater Corros 67(6):639
2. Deutscher Ausschuss für Stahlbeton (DAfStb) (2001) DAfStb-Instandsetzungs-Richtlinie: Schutz und Instandsetzung von Betonbauteilen—Teil 1. Ausgabe Oktober 2001
3. European Committee for Standardization (2015) European Standards EN 206

4. Breit W (1997) Untersuchungen zum kritischen korrosionsauslösenden Chloridgehalt für Stahl in Beton, Schriftenreihe Aachener Beiträge zur Bauforschung—Institut für Bauforschung der RWTH Aachen, vol 8, Aachen, Dissertation
5. Angst U, Elsener B, Larsen CK, Vennesland Ø (2009) Cem Concr Res 39(12):1122
6. Glass GK, Buenfeld NR (1997) Corros Sci 39(5):1001
7. Harnisch J, Raupach M (2010) From the earth depths to space heights. In: Proceedings of EUROCORR, Moscow
8. Angst U, Elsener B (2015) Proceedings of ICCRRR, Leipzig
9. Raman RKS (2006) Eng Fail Anal 13(7):1044
10. Angst U, Elsener B, Larsen CK, Vennesland Ø (2011) Corros Sci 53:1451
11. Garcia V, François R, Carcasses M, Gegout P (2014) Mater Struct 47(9):1483
12. Dauberschmidt C, Vestner S (2010) DBV-Heft 19:65
13. Alonso C, Castellote M, Andrade C (2002) Electrochim Acta 47(22–23):3469
14. Sanchez AN, Sagüés AA (2012) Proceedings of NACE international corrosion conference, Salt Lake City
15. Soylev TA, François R (2003) Cem Concr Res 33(9):1407
16. Deutscher Ausschuss für Stahlbeton (DAfStb) (1989) Arbeitskreis „Prüfverfahren Chlorideindringtiefe“: Anleitung zur Bestimmung des Chlorid-gehaltes von Beton. Berlin: Beuth. In: Schriftenreihe des Deutschen Ausschusses für Stahlbeton, vol 401, p 9

Part III
Behaviour of Corroded Structures

Corrosion-Induced Degradation of Reinforced Concrete Elements: Preliminary Results

O. Loukil, L. Adelaide, V. Bouteiller, M. Quiertant, T. Chaussadent, F. Ragueneau, X. Bourbon and L. Trenty

Abstract Steel reinforcement corrosion is the predominant cause of deterioration of reinforced concrete structures. Corrosion products can induce concrete cracking, loss of adhesion at the steel-concrete interface, loss of reinforcing bar cross-section, and even spalling of the concrete cover. This study investigates the durability problems related to corrosion of the reinforcement by combining experimental and numerical approaches. This paper particularly focuses on the experimental methodology used for the evaluation of the development over time of damages (steel corrosion products formation and crack patterns) induced by accelerated corrosion tests. The accelerated corrosion tests were carried out by applying a constant current between a reinforcement bar used as an anode and an external counter electrode. The corrosion process was monitored by measuring relevant electrochemical parameters, i.e. free corrosion potential, polarization resistance and electrical concrete resistance. The aim of this paper is to characterize the crack pattern (orientation, length and width) and the corrosion products layers according to the current density and the time.

Keywords Reinforced concrete · Corrosion · Rust layer
Cracks · Electrochemistry

O. Loukil (✉) · L. Adelaide · V. Bouteiller
Université Paris-Est, MAST, SDOA, IFSTTAR, F-77447 Marne-la-Vallée, France
e-mail: olfa.loukil@ifsttar.fr

M. Quiertant
Université Paris-Est, MAST, EMMS, IFSTTAR, F-77447 Marne-la-Vallée, France

T. Chaussadent
Université Paris-Est, MAST, CPDM, IFSTTAR, F-77447 Marne-la-Vallée, France

F. Ragueneau
LMT/ENS Cachan/CNRS/Univ. Paris 6/PRES UniverSud, 94230 Cachan, France

X. Bourbon · L. Trenty
Andra, 92298 Châtenay-Malabry Cedex, France

C. Andrade et al. (eds.), *Service Life and Durability of Reinforced Concrete Structures*, RILEM Bookseries 17, https://doi.org/10.1007/978-3-319-90236-4_10

1 Introduction

Steel reinforcement corrosion in Reinforced Concrete (RC) structures of is a major issue of concern for asset owners. The consequences of this phenomenon are degradation of the steel/concrete bond, reduction of the steel reinforcing bar (rebar) cross-section, and concrete cover cracking [1]. This last aspect results from the production of oxides which occupy a volume that is two to seven times higher than the parent steel. Actually, when this production is greater than diffusion of iron oxide into the concrete pores, pressure increases at the interface between the rebar and the surrounding concrete [2–5] and exerts tensile stresses in the concrete all along the corroding reinforcements. If these stresses exceed the concrete tensile strength, cracking initiates and propagates towards the outer surface leading to the delamination of the concrete cover [6, 7]. Due to the deleterious mechanical effect of the corrosion phenomenon, it is important to develop non-destructive techniques as well as predictive numerical modelling to assess the corrosion evolution of RC structures. This could help asset owners to adopt an efficient maintenance policy. The present study focuses on chloride-induced corrosion and the main issue of this paper is to design a specific test protocol to generate a "controlled" corrosion evolution versus time in order to bring some experimental evidences on the concrete cover cracking process due to corrosion and to determine relevant input parameters for the numerical modelling.

2 Experimental Program

2.1 Materials and Specimens

For the experimental campaign, twelve single-rebar specimens (500 × 125 × 100 mm^3) were cast with a 600 mm long and 20 mm diameter steel rebar. This reinforcement bar was positioned to obtain a 30 mm concrete cover at two sides of the specimen (Fig. 1). The specimens designed with an asymmetric location of the rebar aim to get closer to reality. A Portland cement and siliceous aggregates were used for the concrete composition with a water to cement ratio of 0.73. This ratio is representative for former reinforced concrete structures. Moreover, it allows a faster penetration of the chloride ions during the accelerated corrosion test.

The cement type used for the concrete composition was CEM I 52.5 CP2 NF according to European standards. Concrete was prepared with aggregates having different particle size classes [(0/0.315 mm; 0.315/1 mm; 0.5/1 mm; 1/4 mm; 2/4 mm; 4/8 mm; 8/12 mm; 12.5/20 mm)]. Compressive and tensile strengths were measured on concrete cylinders (160 mm in diameter, 320 mm in height) after 28 days according to NF EN 12390-3 [8] and NF EN 12390-6 [9] standards. The mean compressive strength is 32 MPa, the tensile strength is 2.6 MPa, the Young's modulus is 35 GPa and the Poisson's ratio is equal to 0.15.

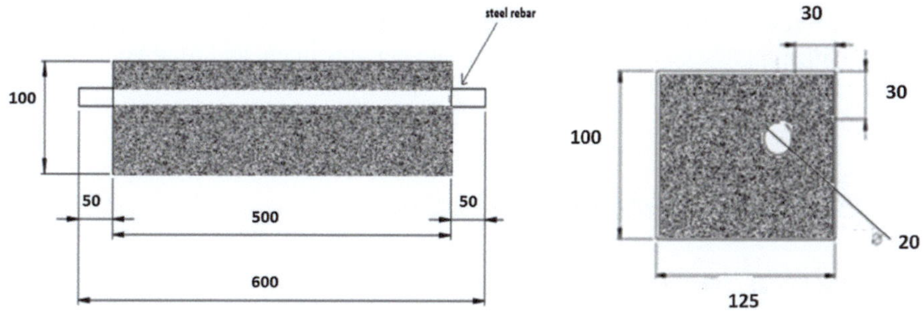

Fig. 1 Schematic representation of RC specimens (dimensions in mm)

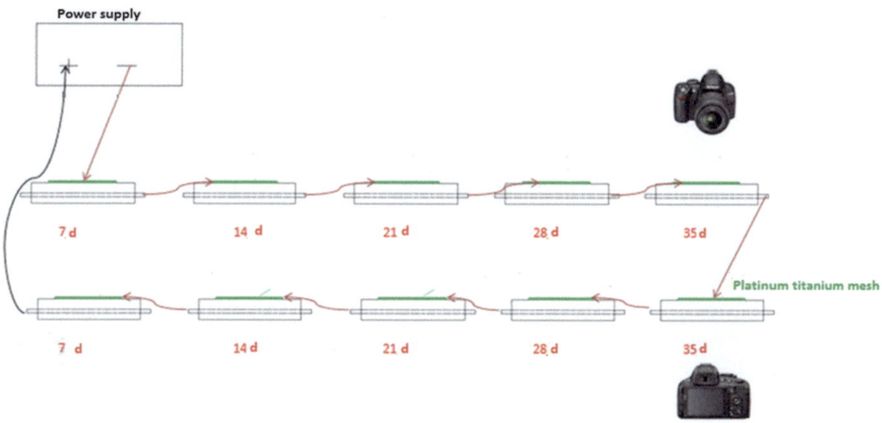

Fig. 2 Scheme of the accelerated corrosion set-up

2.2 Accelerated Corrosion Test and Monitoring Set-up

In this research, experimental investigations have been carried out in the laboratory under accelerated conditions. The set-up used for the accelerated corrosion test and the monitoring system are illustrated in Fig. 2. Corrosion in the RC specimens was promoted by polarizing the embedded steel bars in the anodic direction using a power supply (Agilent 6614 C, 100 V, 0.5 A) which delivered an imposed anodic current. The counter electrode (cathode) was an inert platinum titanium mesh (275 mm long, 75 mm wide) placed into a PVC tank containing an alkaline electrolyte (1 g/L of NaOH, 4.65 g/L of KOH, 30 g/L of NaCl) which was glued on the top side of the concrete (in the middle of the specimen). All specimens were connected in series and a current density of 100 $\mu A/cm^2$ of steel was continually applied during the chosen exposure time (7, 14, 21, 28, and 35 d). At the end of each considered exposure time, two specimens were disconnected from the electrochemical test set-up. One of these two specimens was used for the non-destructive electrochemical characterization and the other one was subjected to destructive measurements. The accelerated corrosion test was monitored using a data acquisition unit (Keysight 34970 A): temperature of ambient air, supplied constant current, voltage per specimen, and total voltage were recorded every two hours.

2.3 Electrochemical Characterizations

Electrochemical characterizations of the rebar were performed after each exposure time of the accelerated test in order to determine the corrosion state. The imposed anodic current was interrupted during the measurements. Half-cell potential measurements (E_{corr}), linear polarization resistance measurements (LPR) and impedance spectroscopy (Re) were carried out using a potentiostat (Bio-Logic, PARSTAT 2263) and the usual electrochemical cell with three electrodes. The working electrode was the steel rebar, the reference electrode was a KCl saturated calomel electrode (SCE, 242 mV/SHE) and the counter electrode was a titanium platinum mesh. The same electrolyte as for the accelerated corrosion test was used.

Then the corrosion current density J_{corr} (μA/cm^2) was calculated based on Eq. (1):

$$J_{corr} = \frac{B}{R_p S} \tag{1}$$

with B a constant (26 mV for corroding steel), R_p (ohm) the linear polarization resistance and S (cm^2) the exposed steel surface (172.78 cm^2 in this study).

2.4 Experimental Observations

2.4.1 Visual Observations

At the end of the accelerated corrosion test, at first non-destructive observations were conducted. Special attention was given to the top and the front sides of specimens because of their lower concrete cover (30 mm). Cracks widths were measured with a crack measuring magnifier (resolution 0.1 mm) on these two surfaces. Then the specimens were sliced (125 × 100 × 20 mm^3) for autopsy, considering corroded areas as illustrated in Fig. 3. After 24 h drying in an oven at 45 °C, the slices were photographed and examined in order to characterize the crack pattern (angle and length) (Fig. 4). The methodology developed by Sanz-Merino [10] was adopted (Fig. 4a) to determine the angular position. The length of the cracks in each cross-section was estimated by superimposing the photographs of the cross-section and a circle graduated every centimeter as shown in Fig. 4b. Then from both results an attempt was made to correlate what would be seen by the asset owner on the concrete surface in 2Dimensions and what happened inside the concrete with an aim of 3Dimensions.

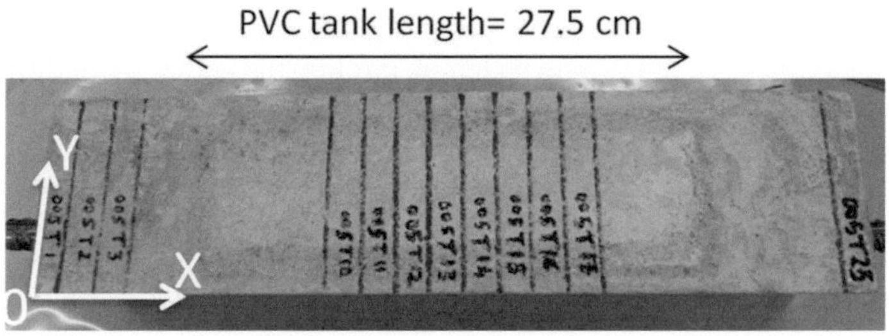

Fig. 3 Graduated circles positioned on the two sides of a slice to measure **a** the crack angular position and **b** the crack length

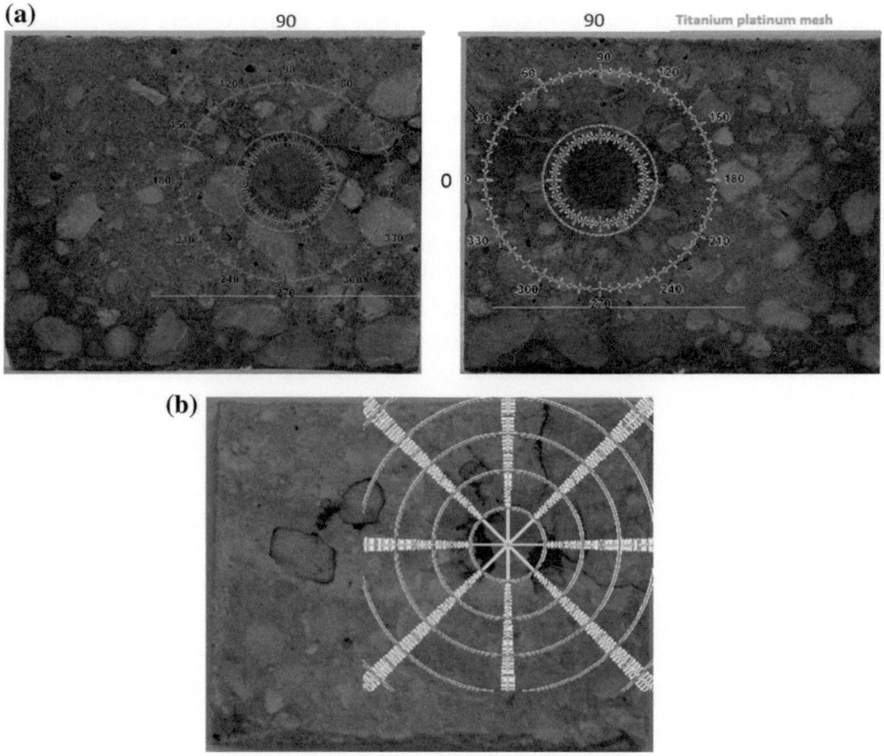

Fig. 4 Determination of the crack angular position (**a**) and the crack length (**b**) after the accelerated corrosion test on the two sections of a slice

2.4.2 Scanning Electron Microscopy Observations

The slices were impregnated with an epoxy resin under vacuum (to prevent disbonding between concrete and steel when the corrosion damage was severe) and then cut into smaller samples which dimensions ($2 \times 4.5 \times 4.5$ cm^3) fitted in the Scanning

Electron Microscopy (SEM) observation room. The first objective is to determine the corrosion product thickness and its length circling the perimeter of the rebar. The second objective is to observe the crack pattern and the transfer of the oxides through the cracks.

3 Experimental Results

3.1 *Voltage Evolution During the Accelerated Corrosion Test*

During the accelerated corrosion test for each specimen the voltage was recorded. Figure 5 shows the time evolution of the voltage for the RC specimens. Before the test (time = 0), the measured voltage reflects the electrical resistance of the specimens which are in a close range from 2.5 to 3 V except for the P004-7d specimen (3.3 V). During the test, three consecutive stages can be distinguished. In the first stage, the observed voltage increase (about 1 V) may be explained by the formation of resistive iron oxides (passive layer) [11, 12] around the rebar and also by the diffusion into the concrete and the filling up of the concrete pores by the oxides in the vicinity of the steel. In the second stage, the drastic voltage decrease (50% loss) can arise from the concrete cracking and the steel/concrete debonding. The last stage demonstrating a constant voltage (1.4 V) appears for longer durations, 28 and 35 d and is likely representative of a constant impedance of the corrosion layer. These preliminary observations need to be discussed considering the migration of the ionic species under galvanostatic control and particularly the penetration of the chloride ions. Analysis of these results is in progress (Fig. 6).

3.2 *Corrosion Rate and Crack Patterns*

Photos of all specimens were periodically taken to record the external changes (cracks and corrosion products) at the front side of specimen, while photographing the top side was only possible at the end of the corrosion process of the considered specimen. Figure 7 shows the corrosion signs and the cracks observed at the top side and the front side of two specimens after a 21 days accelerated corrosion test.

Regarding the top sides, specimen P-007 shows some stains of corrosion products and a visible crack roughly along the steel rebar whereas specimen P-008 is not damaged. However, regarding the front sides, the opposite behaviour is observed: specimen P-007 only exhibits a single spot of corrosion product while specimen P-008 shows significant corrosion products along the crack that follows the rebar. A possible explanation for this difference could be that the aggregate as well as the rebar's ribs are not homogeneously distributed and promote this heterogeneity.

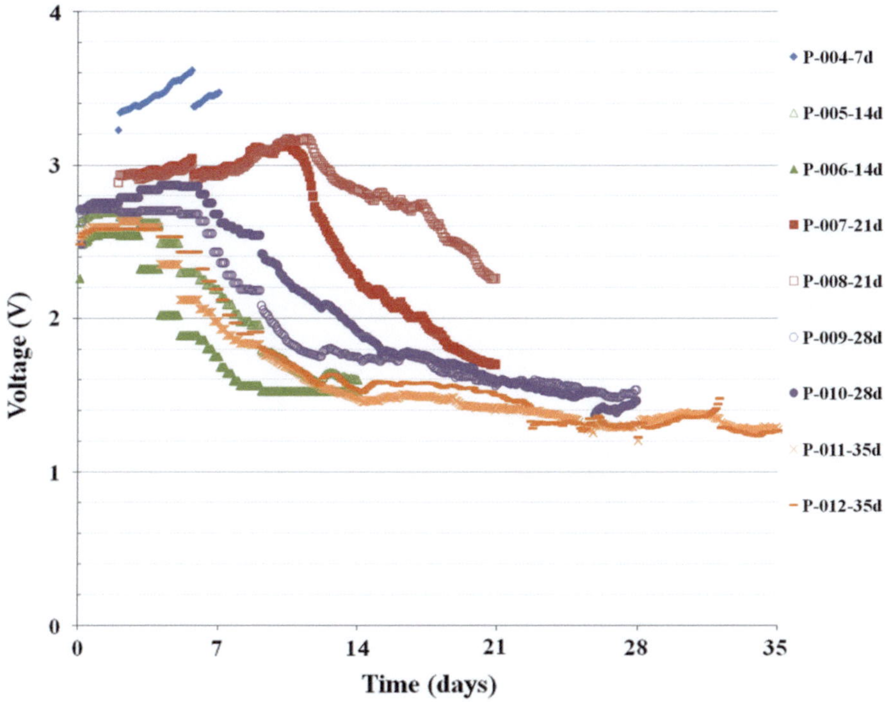

Fig. 5 Voltage evolution during the accelerated corrosion test

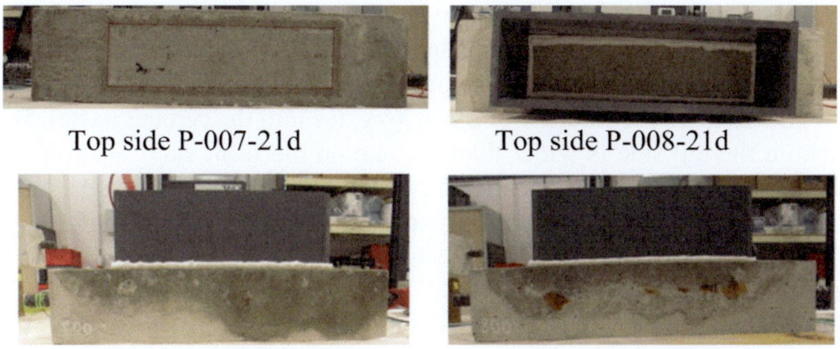

Fig. 6 Qualitative results of the specimens after an accelerated corrosion of 21 days

Measured values of crack width (wider opening measured on top side and wider opening measured on front side) and of J_{corr} are presented in Fig. 7. These results reveal that all corroded specimens for over 14 days have a J_{corr} equal to $10\mu A/cm^2$ approximately except specimen P-004-7d corroded for 7 days. The proposed assumption is that at 7 days chloride ions have not reached the steel/concrete interface yet

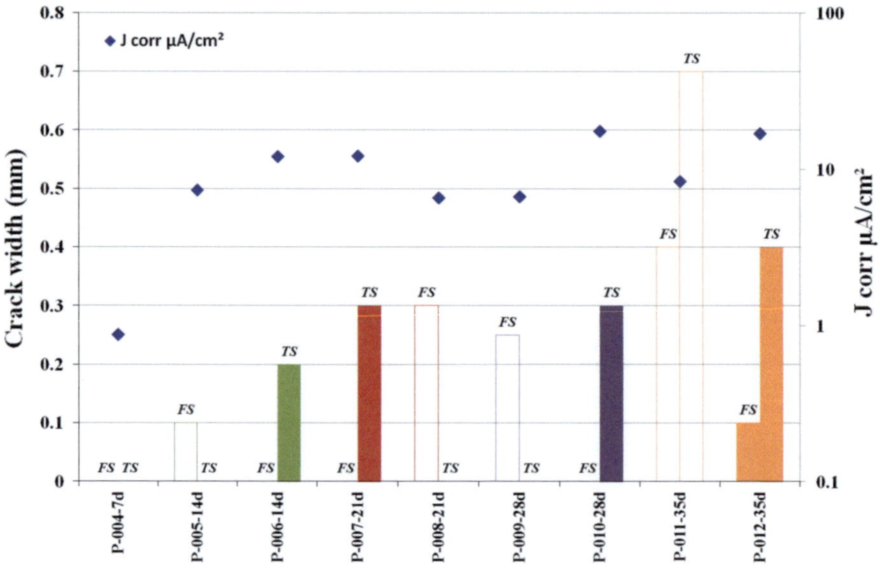

Fig. 7 Crack width evolution and J_{corr} (FS: front side; TS: top side)

and consequently, the steel is still in a passive condition. Measurements of chloride ingress with $AgNO_3$ were made on each slice and the results obtained confirmed this assumption.

Regarding the crack width, the behaviour of corroded specimens appears to be different. As previously mentioned, all specimens are not cracked on the same side. As already suggested, this difference could be ascribed to the heterogeneity of the concrete (the random distribution of the aggregates into the cement paste). Besides, the crack on the top side is wider than the one on the front side. This observation might be attributed to the close localization of the counter electrode on the top side of the specimen (relative to the position of the steel rebar) that permits chloride ions to quickly reach the rebar (modification of the physico-chemical conditions that locally enhances the corrosion process).

Figure 8 shows the evaluation of the crack patterns (angular position (a), length (b) and width (c)) for each of the 16 sides of the eight slices of specimen P-008-21d. The cracks propagate from the steel/concrete interface to the concrete surface.

As shown in Fig. 8a, two groups of the position of the cracks near the steel surface are observed oriented close to 40° and 120° according to the graduated circle (see Fig. 4a). Two mains cracks seem to be displayed. Two sets of crack lengths could be identified, the first one is about 3 cm long (crack extending through the entire concrete cover) and the second one is between 1 and 2 cm long, see Fig. 8b. The lengths tend to fluctuate because of the distribution of the aggregates which influences the crack path. The graph in Fig. 8c corresponds to crack widths and shows the observed general trend. The crack widths are around 0.1 and 0.3 mm.

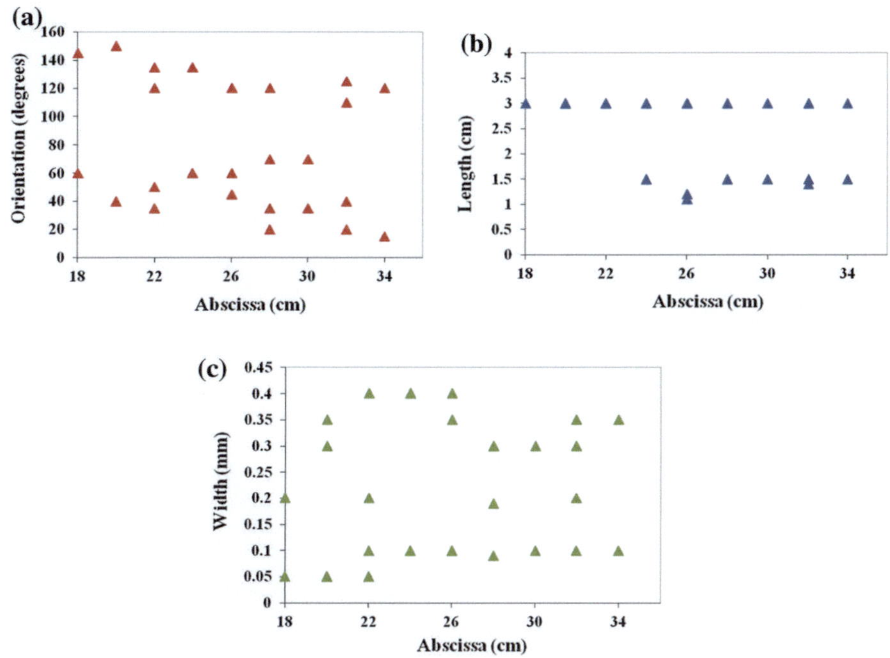

Fig. 8 Measurement results for the characterization of the internal cracks **a** angular position, **b** length and **c** width

The thickness of corrosion products is measured by Scanning Electron Microscopy at different zones along the steel/concrete interface (see Fig. 9) and the perimeter represents the length of the shape of the rebar according to the volume of the corrosion products.

The minimum thickness of corrosion products varies from 50 to 57 μm and the maximum thickness varies from 257 to 314 μm (see Table 1). The highest thickness is located in the upper part of the steel rebar. This 'expected' difference could result from the distance between the counter electrode and the steel surface area. The closer they get, the more the corrosion process is forced. The crack width located on the top side is wider than the one located on the front side because the penetration path of the chloride ions to reach the steel surface area is the shortest.

Comparing the internal measurements 'crack orientations' and the widths measured on the sliced samples with the external observations (corrosion products spots on the front side of the specimens), the surface observations do not reflect the actual internal corrosion state at the steel/concrete interface.

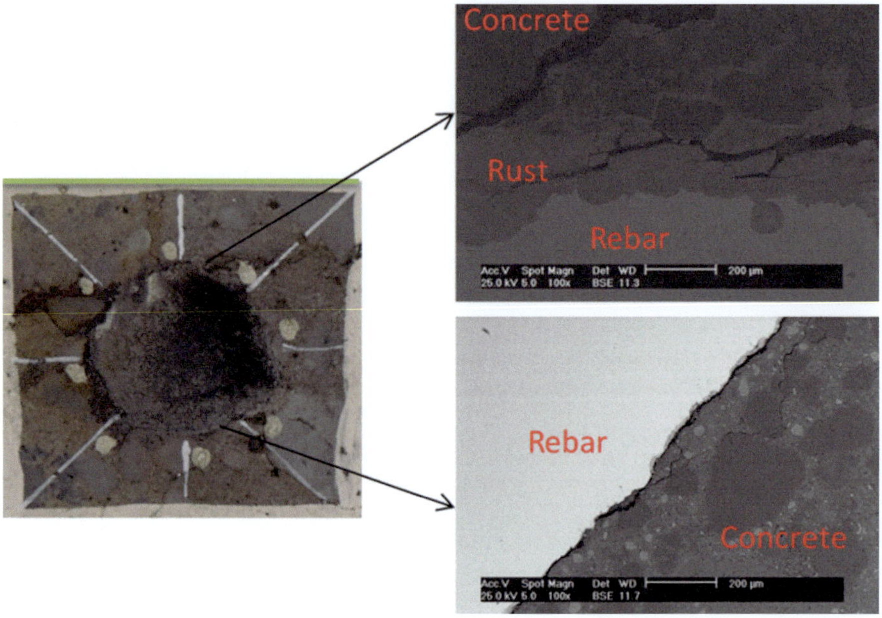

Fig. 9 Methodology for the estimation of the rust layer thickness at different zones of the steel/concrete interface

Table 1 Thickness of the corrosion products layers and corresponding visual distribution of evidences on the front side

Slices	Maximum thickness of the corrosion products layers (μm)	Minimum thickness of the corrosion products layers (μm)	Perimeter of the corrosion products layers Max/Min (μm)
P-008-21d-T10	257	57	2617/5233
P-008-21d-T13	314	63	5233/5233
P-008-21d-T16	257	50	5233/10,467
Mean values	276	56.6	

4 Conclusions

Cracks due to corrosion of the steel reinforcement in concrete specimens have been investigated using an accelerated corrosion test. A specific methodology able to describe the main features of the tests is proposed. Based on the results, the following preliminary conclusions can be drawn:

- Three consecutive stages in the corrosion process can be distinguished: the increase of the polarization resistance may be explained by the development of resistive iron oxides (passivation layer). The second stage could highlight the loss of the resistance of the set due to the concrete cracking and the loss of bond between the steel and concrete surface area, respectively. Regarding the third stage, the observed effect may be attributed to the fact that the properties of corrosion layers remain unchanged. Then, the value of the voltage at the end of the test certainly reflects the resistance of both cracked concrete cover and iron oxide layer.

- After 7 days, an active state of corrosion is not clearly observed and this could be explained by the fact that a sufficient amount of chlorides to initiate corrosion has not reached the steel/concrete interface.
- After 21 days, the two cracks orientations (40°/120°), lengths (3 cm/1.6 cm) and width (0.3 mm/0.1 mm) are coherent for the same specimen.
- From Scanning Electron Microscopy results, corrosion products thicknesses are in the range of 56.6 μm/276 μm.
- The corrosion-product spots on the surface of the concrete specimens do not reflect the internal corrosion state at the steel/concrete interface of the specimen.

Results reported here have to be confirmed after achieving the measurements on all specimens. In the following steps, some of the experimental results such as the thickness and the display of the oxide layer will be used as input data for the numerical modelling of corrosion-induced cracks. The other experimental results associated with the crack patterns will allow a comparison between the experience and the modelling. An experimental test is in progress to characterize mechanical properties of these products to improve the modelling of the corrosion product layer in the numerical simulation.

References

1. Cairns J, Millard S (1999) Reinforcement corrosion and its effect on residual strength of concrete structures. In: Proceedings of the 8th international conference structural faults + repair-99, London
2. Weyers R, Prowell B (1996) Corrosion inhibiting repair and rehabilitation treatment process for reinforced concrete structures. Cem Concr Compos, 459
3. Mehta PK, Monteiro PJM (1997) Concrete: structures, properties and materials. Indian Concrete institute, India
4. Richard B, Quiertant M (2012) Influence of accelerated corrosion on the reinforced cover concrete cracking behavior: experimental and numerical study. Eur J Environ Civil Eng 16(3–4):450–459
5. Richard B, Quiertant M (2016) Experimental and numerical analysis of corrosion-induced cover cracking in reinforced concrete sample. Comput Concr 18(3):421–439
6. Jamali A (2013) Modeling of corrosion-induced concrete cover cracking: a critical analysis. Constr Build Mater 42:225–237
7. Dehoux A (2012) Propriétés mécaniques des couches de produits de corrosion à l'interface acier/béton. Université Pierre et Marie Curie
8. NF EN 12390-6 (2001) Testing hardened concrete. Part 6: tensile splitting strength of test specimens
9. NF EN 12390-3 (2003) Testing hardened concrete. Part 3: compressive strength of test specimens
10. Sanz Merino B (2014) Experimental and numerical study of cracking of concrete due to corrosion. Universidad Politecnica de Madrid Escuela Tecnica Superior de Ingeniros de Caminos, Canales y Puertos, p 254

11. Caré S, Raharinaivo A (2007) Influence of impressed current on the initiation of damage in reinforced mortar due to corrosion of embedded steel, vol 37, pp 1598–1612
12. Poupard O (2004) Corrosion by chlorides in reinforced concrete: determination of chloride concentration threshold by impedance spectroscopy. Cem Concr Res 34:991–1000

Part IV
Corrosion Prevention and Repair Techniques

Development of a Test Method for the Durability of Carbon Textiles Under Anodic Polarisation

Amir Asgharzadeh and Michael Raupach

Abstract Cathodic protection (CP) is a widely used method to protect steel reinforcements against corrosion. In the course of the last half century, it has been established as a proven method for repairing corrosion-affected reinforced concrete structures, exposed mainly to chlorides. The impressed current anode system for the protection of steel in concrete is already known and applied in practice for more than 30 year. The CP anodes can be embedded in mortar, applied as coating or distinct anode on the repair structure surface. Using CP, the potential of the steel reinforcement is shifted in cathodic direction towards more negative values and consequently the anodic dissolution of iron is suppressed. The current densities on the surface of the reinforcement play a key role in shifting of the potential in cathodic direction. For the anode material nowadays Mixed Metal Oxide coated Titanium (MMO) is often used due to its high durability under anodic polarization. As an alternative anode material, Carbon Textiles embedded in mortar, which provide extraordinary mechanical properties and are also conductive, have already been studied. The results demonstrate that Carbon Textiles are generally suitable as an anode material for CP on reinforced concrete. Deterioration of Carbon is likely to occur in CP systems using carbon as anode material, when critical current densities are exceeded. Therefore the durability of Carbon Textiles at typical CP current densities of less than 20 mA/m^2 with respect to concrete surface is considered of special interest.

1 Introduction

In the past decade in certain cases Carbon fibres have been used as anode material for CP on steel reinforced concrete. In comparison to conventional anode materials, such as MMO coated titanium, carbon offers several benefits e.g. reduced costs depending on the type of Carbon Textile, flexible installation, structural strengthening or

A. Asgharzadeh (✉) · M. Raupach
Institute of Building Material Research, RWTH Aachen University, Schinkelstrasse 3, 52062 Aachen, Germany
e-mail: Zadeh@ibac.rwth-aachen.de

C. Andrade et al. (eds.), *Service Life and Durability of Reinforced Concrete Structures*, RILEM Bookseries 17, https://doi.org/10.1007/978-3-319-90236-4_11

limitation of crack widths. Beside these benefits, first investigations of various carbon textiles embedded in mortar specimen under laboratory conditions show promising results concerning the performance of carbon textiles as anode material for CP [10].

The use of Carbon reinforced concrete provides many new opportunities for building and construction industries. In comparison to the traditionally used steel rebars, which need a minimum thickness of concrete cover to provide sufficient corrosion protection, carbon reinforced concrete is not susceptible to corrosion. As a result, it can be used to build very thin and light element structures, while still offering high strength, see e.g. Fig. 1.

However, there are still few investigations about the long term durability of carbon when in direct contact with alkaline environments.

In order to investigate the suitability of carbon textiles as anode material for CP, different experiments focussing on the anodic polarisation behaviour of carbon were performed. So far, the exposure conditions of the experiments were limited to aqueous solutions, simulating the alkaline environment of uncarbonated concrete. In a next step, carbon textiles are embedded in mortar specimens to simulate realistic conditions. In addition, the influence of chlorides on the durability of the textiles, as well as on the polarisation behaviour of carbon textiles, will be further investigated.

2 Literature Review

According to the Pourbaix diagram (Fig. 2), carbon is a not stable material in the pH range of normal concrete at its own free corrosion potential. Instead, carbon will rather dissolve into substances like carbon acid, carbonate and bicarbonate, or it dissolves into gaseous molecules such as CO_2, CO or CH_4 [1]. The same products of reaction can occur due to high anodic polarization (cf. Eqs. (1)–(3)).

In the following, results of different authors are reflected, which agree in principle. In 1999, Eastwood et al. [2] presented results about the anodic reaction of carbon

Fig. 1 Example for a textile concrete structure at the RWTH Aachen

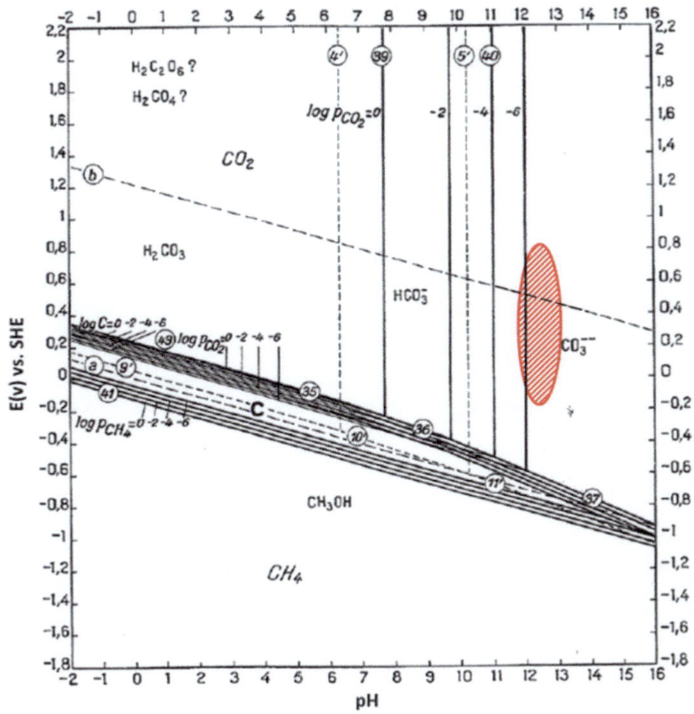

Fig. 2 Pourbaix diagram of graphite

under the presence of carbonate and bicarbonate. They discovered that the anodic reaction(s) occur at much lower anodic potentials compared to solution with the same pH, but without carbonate or bicarbonate. Their investigations also show, that an increased graphite content of the carbon increases the oxidation resistance. Chini carried out some tests regarding the polarization behaviour of carbon in synthetic alkaline solution with pH 13.2 [1]. The results indicate that carbon shows a passive behaviour up to a polarization of 300 mV versus OCP (Open Circuit Potential).

In the case of more positive potentials, carbon acts as a corrosion active material in concrete structures, even though there is a second passive section above 2500 mV versus SCE (Fig. 3). For low pH values, Maaß, as well as Eastwood et al. defined the critical potential where carbon gets electrochemically oxidised with 207 mV versus NHE in acids [2, 3]. According to [4, 5] the anodic reactions of carbon when exposed to acids can be described as follows:

$$C + 2H_2O \rightarrow CO_2 + 4H^+ + 4e^-, \varphi_{00} = 0,207VNHE \tag{1}$$

$$C + H_2O \rightarrow CO + 2H^+ + 2e^-, \varphi_{00} = 0,518VNHE \tag{2}$$

$$CO + H_2O \rightarrow CO_2 + 2H^+ + 2e^-, \varphi_{00} = -0,103VNHE \tag{3}$$

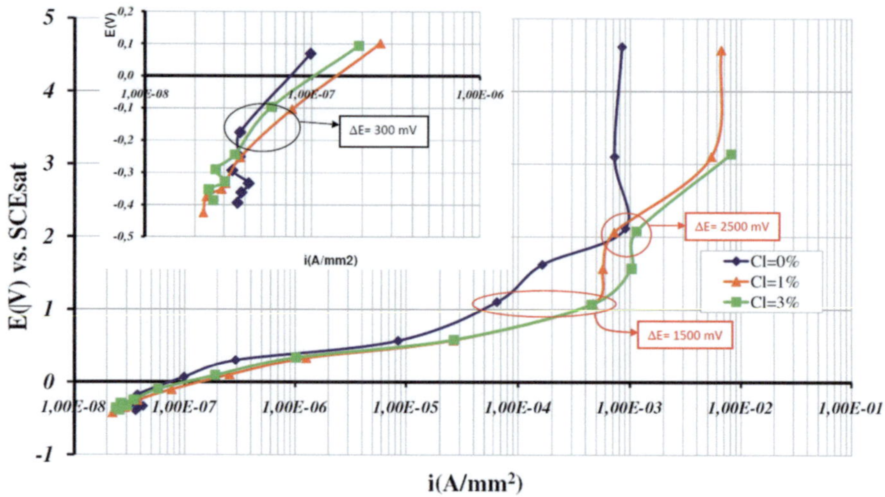

Fig. 3 Current density-potential-curve for carbon in simulated concrete pore solution with different chloride contents [9]

with:

φ_{00} Standard reduction potential of the half-cell reaction at 25 °C, 1013 mbar, pH = 0 (vs NHE)

Since the reaction rate is limited in the range below 207 mV versus NHE due to the kinetics, Maaß stated that the corrosion rates here are very low. Furthermore he described the corrosion rate depending on the applied imposed anodic potential. Corresponding to the Butler–Volmer–Equation, the corrosion rates rise exponentially above the thermodynamic steady-state potential of 203 mV versus NHE [3]:

$$i = i_0\left[\exp\left(\alpha\frac{zF}{RT}\eta_D\right) - \exp\left(-(1-\alpha)\frac{zF}{RT}\eta_D\right)\right] \tag{4}$$

with:

- i Current density [A/m^2]
- i_0 Exchange-current-density [A/m^2]
- α Transfer coefficient [-]
- z Number of electrons transferred [-]
- F Faraday constant [C/mol]
- R Gas constant [J/mol K]
- T Absolute temperature [K]
- η_D Overpotential [V]

Vennesland et al. [6] presented results, which are in conformity with Maaß. By carrying out laboratory experiments they showed, that the electrochemical reactions are changing according to the applied potential. Thus, the potential is one of the key factors affecting the corrosion rate. They proofed their results by measuring the mass loss of carbon after different levels of sustained anodic polarizations. While the carbon specimen polarized with 2 V versus OCP lost 3% of its mass in 168 h testing, the specimen polarized with 3 V versus OCP lost 44%. However, such deterioration rates would be unacceptable for CP-systems, if used on real concrete structures.

A very important factor concerning the boundary conditions seems to be the chloride content of the concrete structure, as it could affect the pH-value of the pore solution as well as the whole binding behaviour of the concrete. This may, as a consequence, lead to a different electrochemical behaviour of the carbon. Here, Chini discovered ambivalent results. While 1 M.-% of mixed-in chloride results in a higher polarisation resistance, 3 M.-% reduced the polarisation resistance of the specimen [1]. Therefore further investigations should be carried out, to investigate the influence of chlorides on the polarisation behaviour of carbon embedded in concrete.

In addition to the factors above, also the temperature, as well as the quality and humidity of the concrete and it´s chemical composition have to be taken into account, when using carbon as anode material for CP for real concrete structures [1].

Furthermore, Raupach et al. performed some test on a CarboCath® area of a parking construction in 2007. Here, the results showed a significantly higher influence of the temperature on the protective current output than of the reduction of the humidity of the test area, as water is consumed during anodic reactions [7].

Pruckner describes in his work the installation process of a carbon anode system for CP. In view of the previous results, a waterproof sealing has been applied to the carbon anodes, as well as a coating on the concrete surface, to reduce the ingress of chlorides [8].

Despite all the existing concerns, Chini rates carbon as generally suitable for use as an anode material for CP, provided that the polarisation (potential shift) of the carbon will remain less than 300 mV versus OCP. For higher potential shifts, the lifetime can be calculated by estimating the mass loss of the carbon, according to Eqs. (5) and (6) [1].

$$\frac{dW}{dt} = \frac{I_{OX} \cdot W_0}{n \cdot F} \tag{5}$$

$$W = \frac{W_0 \cdot Q}{n \cdot F} \tag{6}$$

W Mass loss per unit of time [kg/s]
W_0 Mole mass of carbon [12.01 g/mol]
n Number of the electrons involved in the electrochemical reaction [equiv. mol^{-1}]
F Faraday constant [964853 A sol]
I_{OX} Current because of oxidation [A]

Here, the relation between the mass loss of the carbon and the measured electrical current, as well as the Evan’s diagram, have been used to calculate the life time of the CP system. Chini has shown that a system will theoretically remain operative for

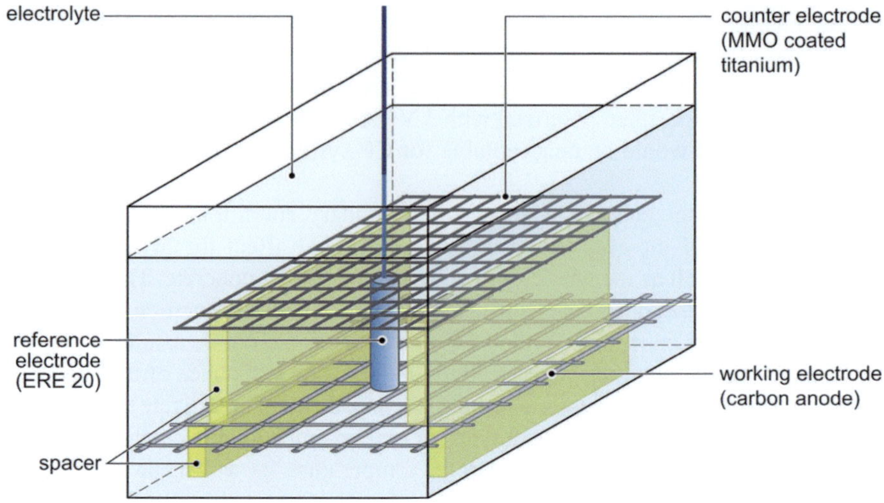

Fig. 4 Schematical test set-up

more than 50 years, if a potential shift between 500 and 750 mV versus SCE is used, which are common values in practice. However, it is important to notice that Chini has only tested the longitudinal properties of the carbon fibers whereas the durability of carbon as textile mesh has not yet been clarified [9].

3 Experimental Program

3.1 General Set-up

Before the durability of carbon textiles embedded in mortar specimen will be tested, basic investigations on Carbon Textiles in synthetic concrete pore solutions were performed. Preliminary pre-tests were executed in order to determine the relevant influencing factors to the measurement results. Figure 4 shows the test set-up for both pre- and main tests in synthetic pore solution schematically.

In order to perform electrochemical investigations an electrochemical cell has been designed. While carbon mesh acts as the working electrode, titanium mesh is used for the counter electrode. Both electrodes are completely immersed in electrolyte solution. In addition a reference electrode is installed to measure the electrode potential of the working electrode. Non-conductive polymer spacers were used to maintain a constant distance between the working and the counter electrode.

3.2 Materials

Table 1 shows the investigated carbon textile as well as other materials that were needed for the Potentiodynamic experiment.

3.3 Specimen Preparation

For the preparation of the working electrodes, carbon textile mesh was cut into pieces with dimensions of 10.5 cm × 5 cm. The electrical connection with the carbon textile was enabled through the use of wire ferrules. A special resin on epoxy basis has been used for the sealing and to eliminate the corrosion of the wire. Figure 5 shows the stepwise preparation of the cable connection and the working electrode.

For the counter electrode MMO coated titanium mesh was cut into elements with similar dimensions as the carbon working electrode. The electrical connection with the mesh cathode was ensured using luster terminals. Spacers were used to maintain the distance between the working and counter electrode as well as to position the reference electrode. Finally the three-electrode cell arrangement was placed in a container having a horizontal cross section of about 15 cm × 8 cm. 630 mg saturated calcium hydroxide solution was required to ensure total covering of the cell arrangement. Figure 6 shows the completed three-electrode-cell. A total number of 24 cell arrangements were prepared in order to investigate the relevant influencing factors (see Table 2).

Table 1 Materials used for electrochemical tests

Element	Material
Working electrode	styrene-butadiene-rubber impregnated Carbon Textile roving distance 20 mm in both directions
Counter electrode	MMO coated titanium
Reference electrode	ERE 20, MnO_2
Electrolyte solution	Saturated $Ca(OH)_2$ Saturated $Ca(OH)_2$ + 3% NaCl
Spacer	Non-conducting synthetic material from PVC

3.4 Testing Program

The Pre-tests were carried out to determine which factors for this test set-up will have a significant influence on the measurement results. Table 2 provides an overview of the factors which were investigated in particular.

The sealing of the roving-tips was achieved by an epoxy-based resin. Figure 7 shows the roving with unsealed as well as with sealed tips.

3.5 Test Procedure

The same measuring cycle was applied for all 24 cell arrangements. The test procedure was implemented as follows:

A Gamry G300 potentiostat in combination with MnO_2-based reference electrodes (type ERE-20) were used for the measurements. First of all, the Open circuit

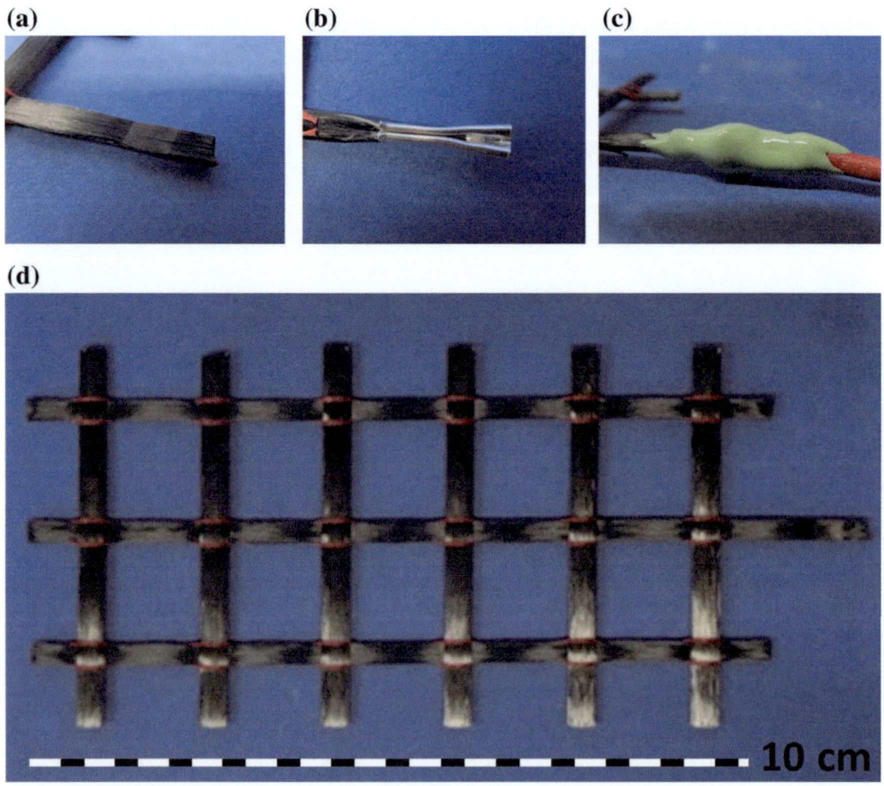

Fig. 5 Preparation of cable connection and working electrode

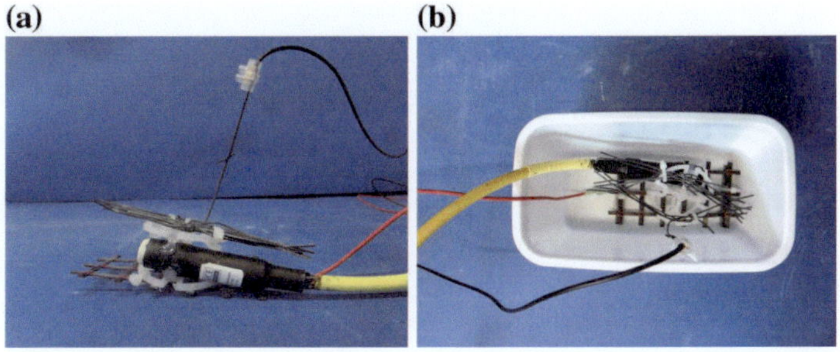

Fig. 6 Completed three-electrode-cell

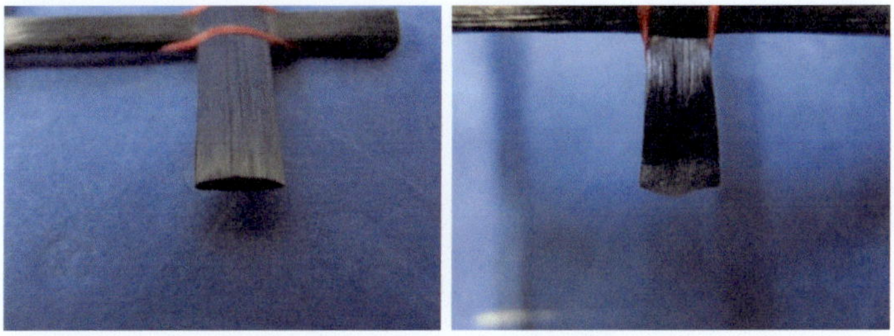

Fig. 7 Unsealed and sealed roving-tip

Table 2 Shows the factor which were investigated in particular

Influencing Factor	Variations	Sample description
Specimen orientation (Figs. 8 and 9)	• Vertical position • Horizontal position	• NV_S_Ca_1/2 • NV_L_Ca_1/2
Sealing of roving-tips (Fig. 7)	• Sealed • Unsealed	• V_L_Ca_1/2 • NV_L_Ca_1/2
Electrolyte solution	• Saturated $Ca(OH)_2$ • Saturated $Ca(OH)_2$ + 3% NaCl	• NV_S_Ca_1/2 • NV_S_Ca+3Cl_1/2
Storing of specimens	• Exposure to solution 24 h before test procedure • No pre-exposure to solution	• NV_L_Ca_24_1/2 • NV_L_Ca_1/2
Homogenization-level of solution	• Permanent stirring of solution • Solution without motion	• NV_S_Ca_G_1/2 • NV_S_Ca_NG_1/2

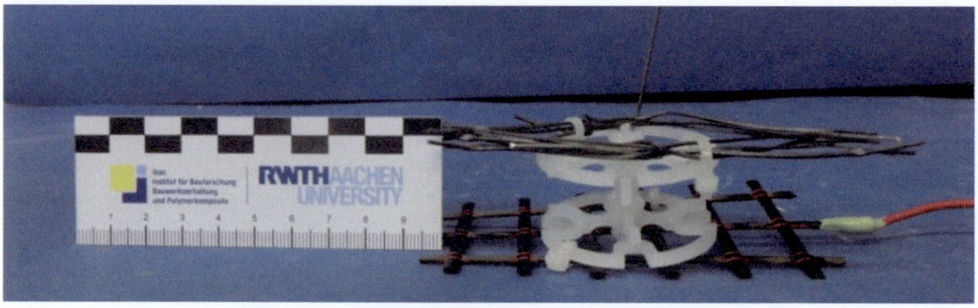

Fig. 8 Side view of a prefabricated cell arrangement

Fig. 9 Top view of a prefabricated cell arrangement

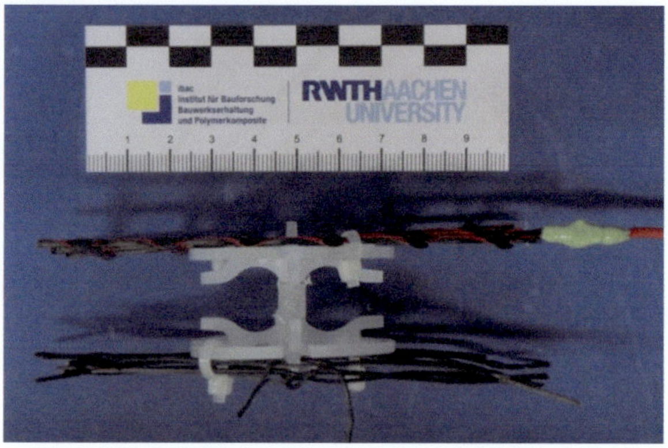

potential (OCP) was measured as an indication for the free corrosion state. For this purpose data points were recorded every five seconds over a total duration of 200 s. Subsequently potentiodynamic polarization with constantly increasing electrical potential was carried out. Starting with the open circuit potential, the applied potential was increased at a constant rate of 0,033 mV per second until an overall potential shift of 2000 mV was achieved. Hence, this resulted in a test duration of approximately 18 h. Automatic data collection took place every 300 s. Finally Instant-off potential measurements were performed to ensure that IR-drop compensated potentials were measured. Overrating of the actual polarization of the anode material can thus be avoided.

Following this anodic polarization, the obtained results were plotted with current-density versus (compensated) potential in order to evaluate the performed tests. The compensated potential can be calculated according to Ohm's law (Eqs. 7 and 8) as follows:

$$U_{compensated} = U_{measured} - \Delta U_{IR} \tag{7}$$

$$\Delta U_{IR} = R \cdot I \tag{8}$$

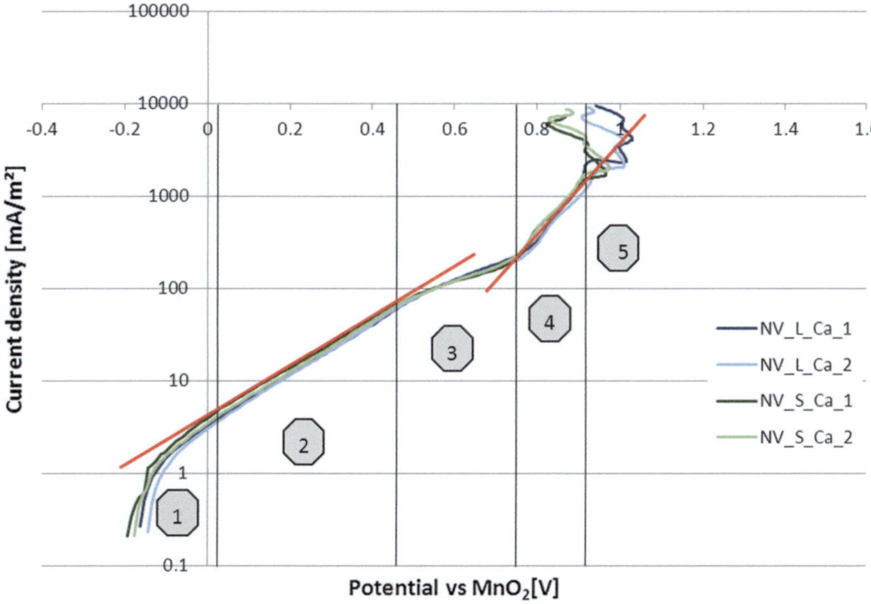

Fig. 10 Influence of specimen orientation

Substracting the voltage (ΔU_{IR}) resulting from the IR-drop ensures that the actual net polarization of the anode material is taken into account.

4 Test Results

For each influencing factor two identical cell arrangements were investigated and the results were compared to their counterparts. Figures 10, 11, 12, 13 and 14 show the test results obtained for both cell arrangements. Note that the current density in the graphs is plotted on a logarithmic scale.

All measured curves show similar typical characteristics, which can be divided into five phases:

1. OCP to Tafel area 1
2. Tafel area 1
3. Transition area
4. Tafel area 2
5. Above Tafel area 2

It should be noted, that for CP-systems in reinforced concrete structures the current density is typically not higher than around 20 mA/m^2 with respect to the concrete surface area. This means, that only phases 1 and 2 are of practical interest.

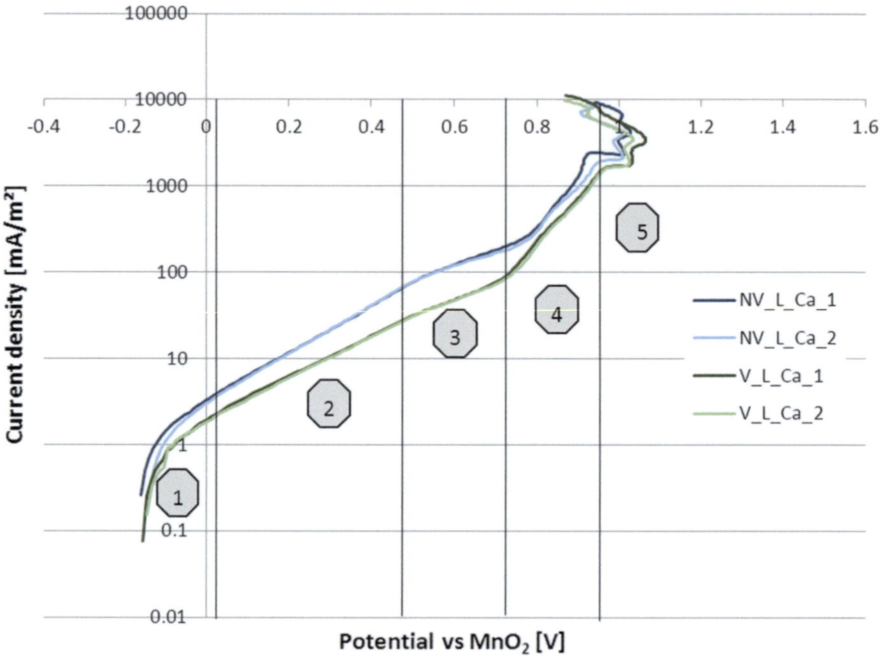

Fig. 11 Influence of sealing of specimen-tips

Figure 10 visualizes the influence of the orientation of the cell arrangement. Since all curves show a very good agreement, the position of the cell arrangements obviously has no significant influence on the measurement results. Conversely, the sealing of roving tips, the storing of the cell arrangements and the homogenization level of electrolyte solution show significant influences on the test results as will be described in the following.

Figure 11 shows the current-density potential curves for the influencing factor *sealing of roving-tips*. A significant gap between the development of the curve for cell arrangements with sealed tips on the one hand (V_L_Ca_1 and V_L_Ca_2) and for cell arrangements without sealed tips (NV_L_Ca_1 and NV_L_Ca_2) on the other hand can be observed. If the same potential is imposed, cell arrangements with no sealing of roving-tips allow a significantly higher current density than cell arrangements with sealed roving-tips. Thus sealing of roving-tips strongly influences the test results, because the sealing of roving tips prevents the passage of electrical current into solution.

Similarly the difference in storing of cell arrangements shows significant effects on the measurement results, as can be seen in Fig. 12. Pre-exposure to electrolyte solution 24 h before the polarization test-procedure leads to an increased current flow during the electrochemical investigations.

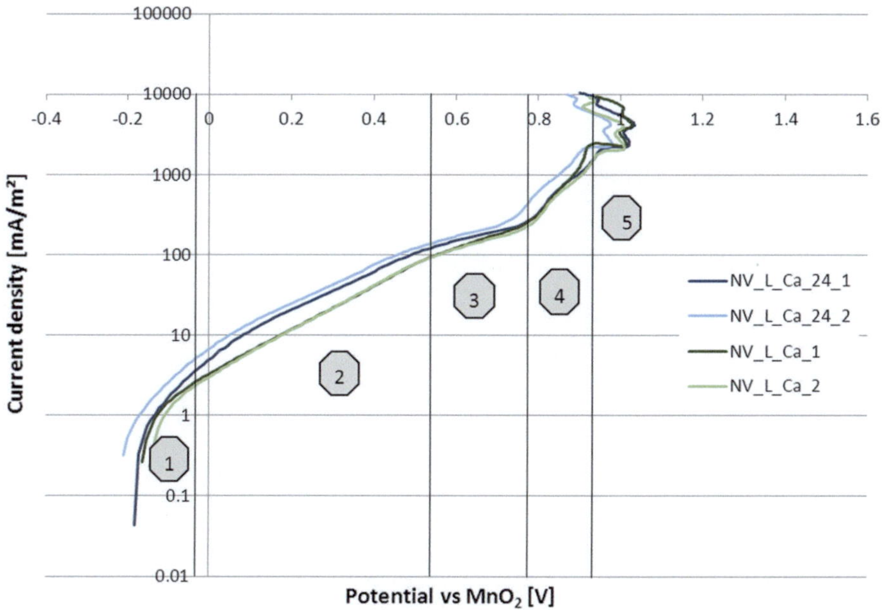

Fig. 12 Influence of specimen storing

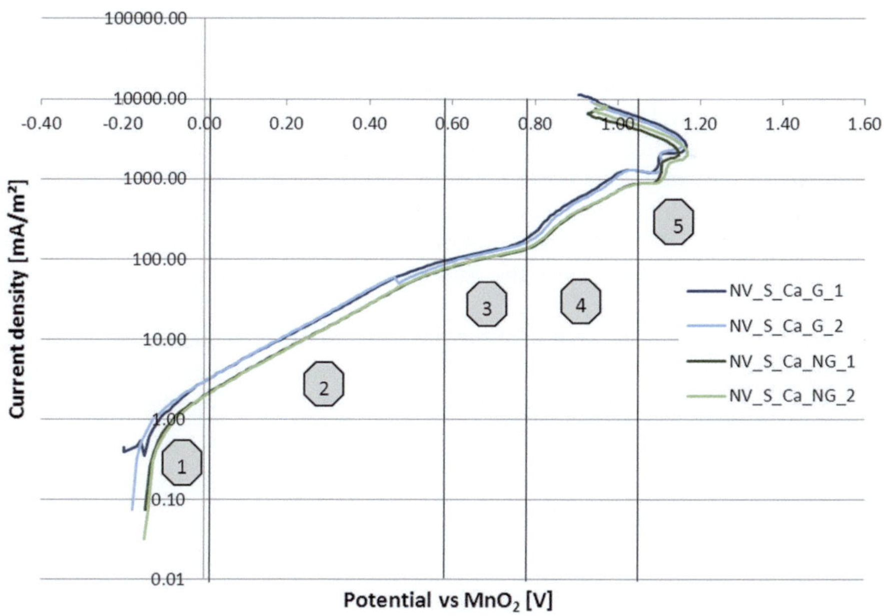

Fig. 13 Influence of homogenization-level of electrolyte solution

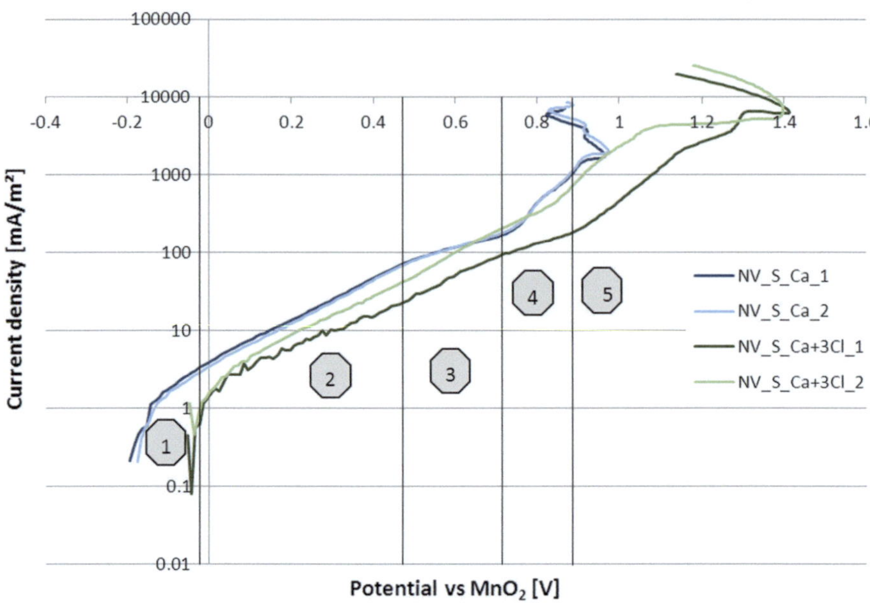

Fig. 14 Influence of electrolyte solution

Figure 13 shows the current-density potential curve with variations concerning the homogenization level of the electrolyte solution. The stirring of the solution shows an impact on the development of the polarization curve as well. The approximation of the curve representing specimen NV_S_Ca_G_2 to the curves representing the specimens in unstirred solution (NV_S_Ca_NG_1 and NV_S_Ca_NG_2) can be explained by the technical impact of the stirring process.

The influence of the electrolyte solution, which is presented in Fig. 14 has to be investigated in further tests since the data obtained from these tests does not have sufficient validity.

Table 3 summarizes the results of the pre-tests. In order to qualify the influence as being significant, a potential criterion was established: A potential shift of more than 50 mV at a current density of 20 mA/m^2 was graded as a significant change.

5 Conclusions and Outlook

A test method has been developed to investigate the durability of Carbon Textiles under anodic polarization. In a first test series the influence of relevant factors regarding the testing procedure in a simulated concrete pore solution has been examined. For the main experiments the following factors are taken into account:

- Horizontal position of the carbon mesh

Table 3 Summary of the pre-test results

Influencing factor	Variations	Significant influence
Specimen orientation	• Vertical position • Horizontal position	No
Tips of carbon rovings	• Sealed • Unsealed	Yes
Electrolyte solution	• $Ca(OH)_2$ • $Ca(OH)_2$ +3% NaCl	To be investigated
Storing of specimens	• Exposure to solution 24 h before test procedure • No pre-exposure to solution	Yes
Homogenization-level of solution	• Permanent stirring of solution • Solution without motion	Yes

- Sealed tips of the carbon roving
- Storing of the cell arrangement 24 h before test procedure starting
- Permanent stirring of exposure solution.

Looking at the anodic polarization curves obtained for the carbon mesh, two different sections can be detected at each curve. The two different slopes indicate a different electrochemical reaction of carbon in each of these two potential regions. These different slopes have to be considered in more detail in a future work.

After evaluation of the main test series in simulated concrete pore solution Carbon Textiles will be applied to concrete specimens by casting them into cement-based mortar. Then, long-time experiments will be carried out in order to finally gain information on the long-term durability as anode material for CP. The objective for such long-term experiments is to find out how the durability of carbon under real operating conditions will develop over time.

References

1. Chini M, Antonsen R, Vennesland, Ø, Arntsen B, Mork JH (2013) A review of utilization of carbon fibres as an anodic material in cathodic protection. In: International RILEM workshop on integral service life modelling of concrete structures, p 395–402
2. Eastwood BJ, Christensen PA, Armstrong RD, Bates NR (1999) Electrochemical oxidation of a carbon. Black loaded polymer electrode in aqueous electrolytes. Journal of Solid State Electrochemistry 3(4):179–186

3. Maaß S (2007) Langzeitstabilität der Kathoden-Katalysatorenschicht in Polymerelektrolyt-Brennstoffzellen. Dissertation, Institut für Chemische Verfahrenstechnik der Universität Stuttgart, Göttingen (in German)
4. Pourbaix M (1974) Atlas of electrochemical equilibria in aqueous solutions. Nat. Assoc. of Corrosion Engineers, Houston, Texas, Chapter IV, 17.1
5. Kinoshita K (1988) Carbon, electrochemical and physicochemical properties. Wiley & Sons, New York
6. Vennesland Ø, Haug R, Mork JH (2005) Cathodic protection of reinforced concrete—a system with woven carbon mesh. In: International conference on concrete repair, rehabilitation and retrofitting (ICCRRR), Cape Town, South Africa
7. Raupach M, Bruns M, Wolff L (2007) Installation and investigation of a field test area with the carbo cath CP-System and different surface coatings at the parking Garage Scheibenstrasse in Düsseldorf. Aachen, Germany
8. Pruckner F (2006) New anode for cathodic protection of car parks. Concr Eng Int 10(1):54–57
9. Mork JH, Mayer S, Rosebom K, Tuveson-Carlstrom L, Sederholm B, Sandberg B (2006) Cathodic protection of concrete structure with a carbon fibre mesh anode. In: EUROCORR 2006, Maastricht
10. Chini Mahdi (2010) Pan-based carbon fibre as anode material in cathodic protection systems for concrete structures. Doctoral Thesis, Faculty of Engineering Science and Technology, Department of Structural Engineering of the Norwegian University of Science and Technology
11. Asgharzadeh A, Raupach M, Koch D (2015) Investigations on the suitability of technical textiles for cathodic corrosion protection: Untersuchung zur Eignung von technischen Textilien für den kathodischen Korrosionsschutz. In: Proceedings of the 4th International Conference on Concrete Repair, Rehabilitation and Retrofitting (ICCRRR), Leipzig, Germany

Corrosion Processes of Carbonated Chloride-Contaminated Reinforced Concrete and Electrochemical Chloride Extraction (ECE) Efficiency

Yolaine Tissier, Véronique Bouteiller, Elisabeth Marie-Victoire, Suzanne Joiret and Thierry Chaussadent

Abstract Civil engineering structures and historical buildings can suffer from corrosion of the embedded reinforcing steel once the concrete cover is totally carbonated and/or when chloride ions have reached the steel/concrete interface. In practice, these two types of contamination can be encountered separately or combined requiring implementation of proper repair methods. In this research, carbonated and chloride-contaminated reinforced concrete specimens were studied by three main analysis means: electrochemical characterizations, Raman microspectrometry and scanning electron microscopy (SEM). The specimens were contaminated in a two-step process involving addition of chlorides followed by carbonation of the complete cover. Electrochemical chloride extraction (ECE) was then performed as a repair treatment (1 A/m^2 of steel surface during 8 weeks). The efficiency of the treatment and its impact on the steel/concrete interface and cement matrix were studied during the treatment, after steel depolarization and on the long term (several months) in order to evaluate the durability of the treatment. Electrochemical characterizations showed an increase of corrosion rates comparing the specimens after fabrication and after carbonation (from a negligible level of 0.1 μA/cm^2 to values in excess of 10 μA/cm^2). SEM observations confirmed this significant increase with the identification of a corrosion layer on most of the steel/concrete interface after carbonation. The ECE efficiency was evidenced by a decrease of chloride content below the practical threshold value of 0.4% by weight of cement after a two weeks treatment. Simultaneously a realkalisation ring was observed around the reinforcement bar having a diameter of

Y. Tissier (✉) · V. Bouteiller
Université Paris-Est, MAST, SDOA, IFSTTAR, Marne-La-Vallée F-77447, France
e-mail: yolaine.tissier@ifsttar.fr

Y. Tissier · E. Marie-Victoire
Laboratoire de Recherche des Monuments Historiques, 77420 Champs-sur-Marne, France

Y. Tissier · S. Joiret
Université Pierre et Marie Curie, LISE, UMR 8235,
CNRS, 75252 Cedex 05 Paris, France

T. Chaussadent
Université Paris-Est, MAST, CPDM, IFSTTAR, Marne-La-Vallée
77447, France

C. Andrade et al. (eds.), *Service Life and Durability of Reinforced Concrete Structures*, RILEM Bookseries 17, https://doi.org/10.1007/978-3-319-90236-4_12

about 1 cm after four weeks. Results obtained after depolarization showed that a two weeks ECE treatment allowed the stabilization of the corrosion state of the rebar.

1 Introduction

Complete cover carbonation and chloride ions penetration until the steel/concrete interface are the two main causes of reinforcement corrosion that will lead to concrete degradation. These two types of contamination can be encountered separately or combined in civil engineering structures as well as in historical buildings requiring implementation of proper repair methods.

This paper aims to characterize the corrosion processes of reinforced concretes contaminated by chloride ions and carbonation prior and after repair using an electrochemical chloride extraction (ECE) treatment. For this purpose, four series of specimens were cast based on two cements and two aging processes.

In this paper, the study's complete protocol for a given series is presented (preliminary results can be found in the MSSCE 2016 conference [1]). Results obtained during the aging and during the treatment are presented and discussed. SEM-EDS analyses and Raman microspectrometry were used to characterize the corrosion products layer at the steel/concrete interface. SEM-EDS provided information on the distribution of corrosion products, morphology and elements of composition while Raman microspectrometry allowed the identification of crystallized corrosion phases formed. Electrochemical characterization (non-destructive-testing (NDT)) was used to follow the evolution of the rebar's corrosion state (active/passive) as well as the durability effect of the treatment.

2 Methodology

In this section the features of the specimens and the experimental program established to follow the evolution of the rebar's corrosion state during the aging and during the electrochemical treatment are presented.

2.1 Specimens Formulation and Design

Specimens were reinforced concrete cylinders of 10 cm in height and 4 cm in diameter (Fig. 1). Composition as given in Table 1 was previously used in literature [2] in order to stay close to the granular distribution of a concrete on a micro-concrete scale. The 0.65 water/cement ratio was chosen to improve the chemical reactions within the concrete cover and was representative of old structures.

Two types of cement were selected:

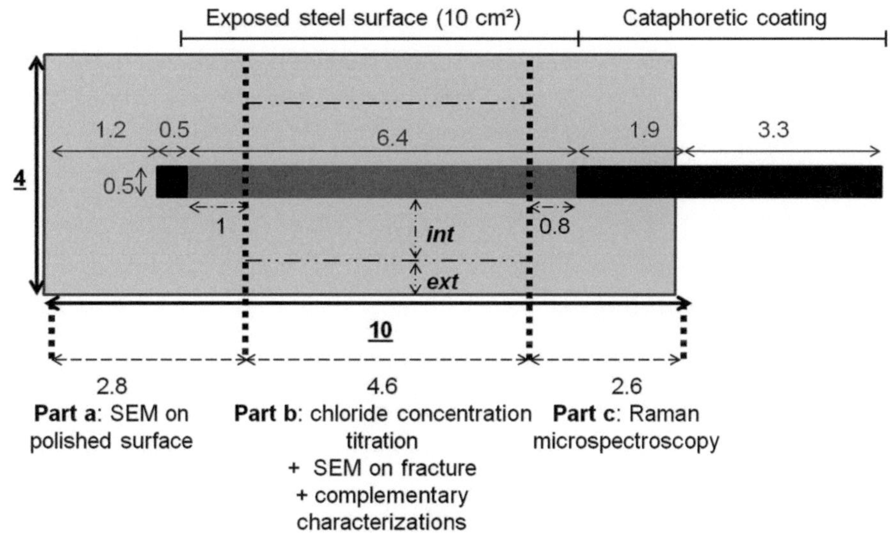

Fig. 1 Specimens design (dimensions in cm) and decomposition in three parts for the analytical characterizations

Table 1 "Micro-concrete" composition

Siliceous Palvadeau aggregates (kg/m^3)					Cement (kg/m^3)	Water (kg/m^3)	w/c ratio
0/0.315	0.315/1	0.5/1	1/4	2/4	387	251.55	0.65
159	295	192	686	229			

- CEM I (52.5 N CP2; Holcim, Lumbres factory) is a usual Portland cement which will allow comparisons with literature results. Specimens cast with CEM I cement were named type 1.
- CEM III/A (52.5 L PM ES CP1 NF; Lafarge, Le Havre factory), is a cement that has been frequently encountered in historical concrete monuments. Furthermore, this cement which contains blast furnace slag is promoted for the construction of new structures because of its lower impact on environmental issue (less CO_2 production). Therefore, the results will help to improve the knowledge of reinforced structures using this cement. Specimens cast with CEM III/A cement were named type 3.

The water accessible porosity measured after curing (28 days for CEM I cement and 90 days for CEM III cement) was about $18\% \pm 1\%$.

The rebar used was a pretreated (polished, cleaned, degreased) smooth round steel type S 235JR of 5 mm in diameter with an exposed surface area (in contact with the surrounding concrete) of 10 cm^2.

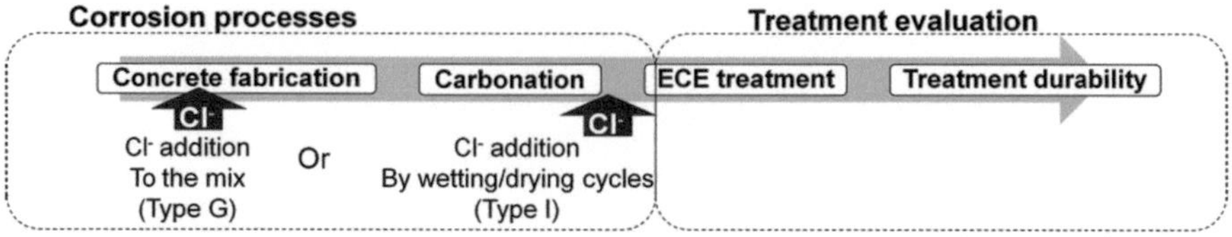

Fig. 2 Experimental set-up

2.2 *Experimental Set-up*

A specific set-up was designed (Fig. 2) in order to study, in a first step, the corrosion of steel embedded in a carbonated and chloride-contaminated concrete and, in a second step, to investigate the efficiency and impact of the electrochemical repair (during the treatment and after its end).

Two different aging procedures were studied:

- Chloride addition to the fresh mix (5% of NaCl by cement weight corresponding to a chloride content of 3% by cement weight) and then, after hardening, carbonation (50% CO_2; 70%RH) until complete carbonation of the concrete cover. These aged specimens were named type G.
- Carbonation until complete carbonation of the concrete cover (50% CO_2; 70%RH) followed by chloride penetration through consecutive wetting/drying cycles (five days submerged in a solution of 35 g/L NaCl followed by two days drying in the laboratory at room temperature and relative humidity). Specimens that were aged according to this procedure were named type I.

Colour indicators were used to determine the carbonated concrete thickness (phenolphthalein (0.5% in ethanol)) and the chloride penetration (0.1 N silver nitrate solution).

After the aging processes, the electrochemical chloride extraction treatment was conducted on 37 specimens of each series (G1, I1, G3 and I3). The set-up of the treatment is presented in Fig. 3. Complying with the NF EN 14038 part 2 standard [3], a current density of 1 A/m^2 of steel was continually applied between the rebar and an external titanium/platinum grid (the counter-electrode was located around the specimens) during 8 weeks (Fig. 3). Electrolyte was a disodium tetraborate solution (25 g/L of $Na_2B_4O_7$ $10H_2O$) which promotes alkaline buffering and chloride entrapment. ECE treatment monitoring relied on room temperature, applied current, voltages (each specimen and all the specimens) records on a one hour frequency base. After 3, 7, 14, 28 and 56 days of treatment, for each series, three control specimens were taken for analyses (Fig. 1). These durations were selected as several authors observed that the extraction of chloride ions was more efficient at the beginning of the treatment [4–7]. The electrolyte was renewed at each of the 5 terms (except at 3 days) to avoid a loss of treatment efficiency due to the increase of chloride ions concentration into the electrolyte.

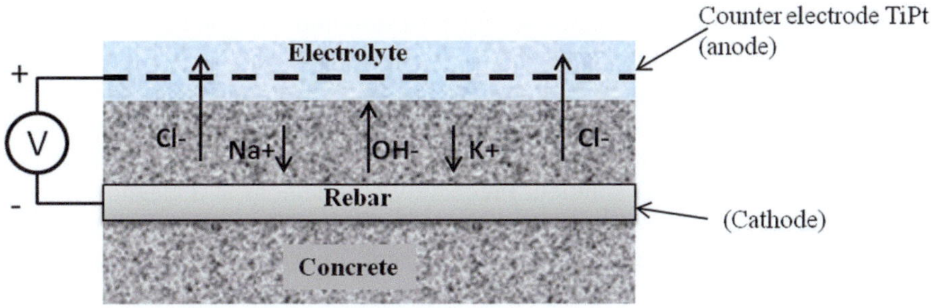

Fig. 3 Set-up of electrochemical chloride extraction

Regarding the chloride extraction evolution, both the free and total chloride concrete content were measured on two of the three specimens by $AgNO_3$ potentiometric titration (at two locations: in the first centimetre around the rebar and in the 0.75 cm of external cover). Moreover the electrolyte chloride titration was also performed. Phenolphthalein test was performed on the surface of part c (Fig. 1) of the same specimens. The third specimen was kept to perform an electrochemical characterization after depolarisation.

After treatment, the electrochemical characteristics of the reinforced specimens were checked every month to evaluate the depolarisation of the rebars (storage at 60%RH and 20 °C) and the stability of the rebars' corrosion state. On a longer term, electrochemical characteristics were followed to ensure the treatment durability (the specimens were then exposed to a temperature of 20 °C and cycles of relative humidity (1 week 92%RH/2 weeks 60%RH).

Simultaneously to the treatment, two specimens were used to observe the occurrence of a lixiviation phenomenon. This was accomplished by immersing both specimens in the tetraborate solution, however without subjecting them to the current. After 28 days, Cl^- content in concrete was analysed on one specimen. After 56 days, the corrosion of the rebar was examined by electrochemical measurements.

2.3 Analysis Means

To study the corrosion processes of the rebar, three main analysis methods were used on different parts of the cylindrical specimen (Fig. 1):

- A Raman spectrometer (LABRAM from Jobin-Yvon Horiba) was used to determine the nature of the formed products. Spectra were obtained with a helium-neon red laser (0.1 mW). A ×50 objective was used to focus laser light and collect spectra for a duration of 600 s. Specific designed cells were prepared to observe in situ (steel/concrete interface isolated from a direct contact with the atmospheric

oxygen) the corrosion processes during corrosion initiation, and during the ECE treatment.
After cutting part c as presented in Fig. 1, a thin glass slide was pasted on the polished surface to prevent the transformation of the corrosion products induced by atmospheric oxygen. After preparation, these samples were immersed in a 25 g/L $Na_2B_4O_7$ $10H_2O$ solution. Frequently observations at a magnification of 50 were made until the detection of pitting corrosion. Raman spectra were then acquired in different areas of interest. The same areas were followed during the ECE treatment and after treatment.

- A scanning electron microscope (SEM JEOL 5600LV) equipped with a micro-analyzer X in dispersal of energy Oxford Link Pentaflet was used to observe corrosion at the concrete/steel interface. Two types of samples were examined: polished cross sections to evaluate corrosion layers evolution and fractured samples to observe both corrosion crystallised phases' morphology and eventual modifications of the cement matrix.
- Electrochemical characterizations were performed with a potentiostat (Bio-Logic, VMP2Z, 5 channels) at each step of the process. The half-cell potential (E_{corr}), the linear polarization resistance (LPR) and the electrical concrete resistance (Re) were measured and the corrosion rate i_{Corr} ($\mu A/cm^2$) was calculated according to Eq. (1).

$$i_{Corr} = \frac{B}{R_p S} \tag{1}$$

With: B = 26 mV, R_p (ohm) the linear polarization resistance and S (cm^2) the active steel surface (10 cm^2 in this study)

This paper focuses on G1 specimens (CEM I cement, NaCl addition to the mix and carbonation). In the first part, corrosion processes are reported. In the second part the ECE treatment efficiency is demonstrated.

3 Corrosion Processes

3.1 Corrosion Products Characterization

Raman microspectrometry. Trials of corrosion followed in situ were made with a G1 specimen before carbonation. Chlorinated green rust (Fig. 4) was the predominant phase found (main Raman bands at 434 and 506 cm^{-1} and a chloride presence characteristic band around 320 cm^{-1}) [8] which evidenced the involvement of chloride into the corrosion processes. Moreover, green rust oxidation into FeIII oxyhydroxide did not occur which proved the good insulation of the interface from a direct exposure to oxygen from the air.

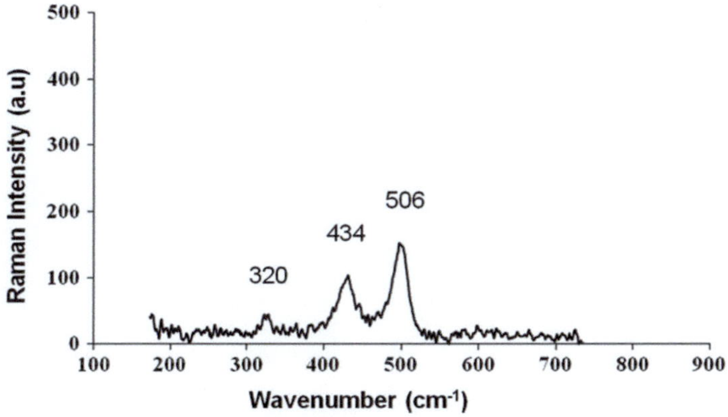

Fig. 4 Chlorinated green rust Raman spectrum

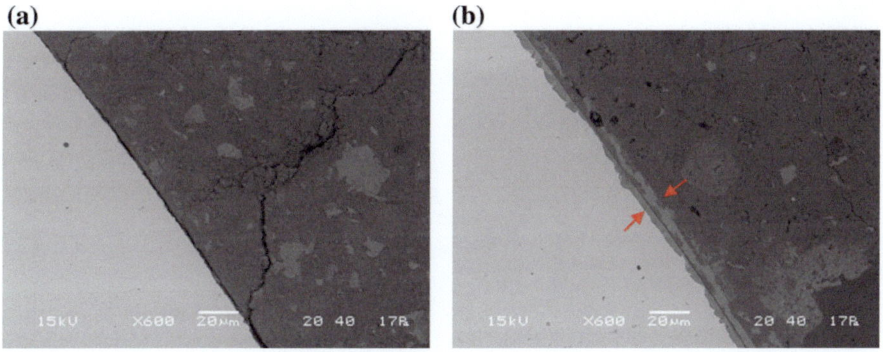

Fig. 5 BEIC-SEM images (×600) of the steel/concrete interface on cross section of specimens type G1, **a** after casting without evidence of corrosion and **b** after carbonation with a thin layer of corrosion products

Scanning electron microscopy. After casting, the steel/concrete interface was mostly characterized by an absence of corrosion (Fig. 5a). Figure 5b shows a representative area of the interface after carbonation with about 5 μm of corrosion products on the steel and 5 μm of migration within the concrete. After carbonation, the corrosion layer thickness could vary from 5 to 20 μm and the penetration of corrosion products within the concrete was mostly of 5 to 15 μm. However on some localized areas a 250 μm thickness was detected. The observation on fracture after carbonation (Fig. 6a) showed numerous areas of corrosion in the rebar imprint in concrete. Figure 6b presents the crystallisation morphology of corrosion products mainly containing Fe, O and Cl. Figure 7 shows a representative result of EDS elemental analysis obtained on corrosion products which highlights the presence of chloride element.

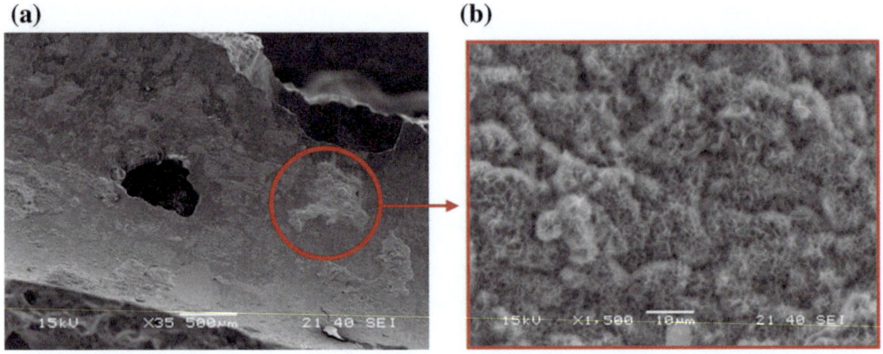

Fig. 6 SEI-SEM images of corrosion products observed into the rebar imprint after carbonation **a** ×35 and b ×1500

Fig. 7 EDS elemental analysis of corrosion products after carbonation

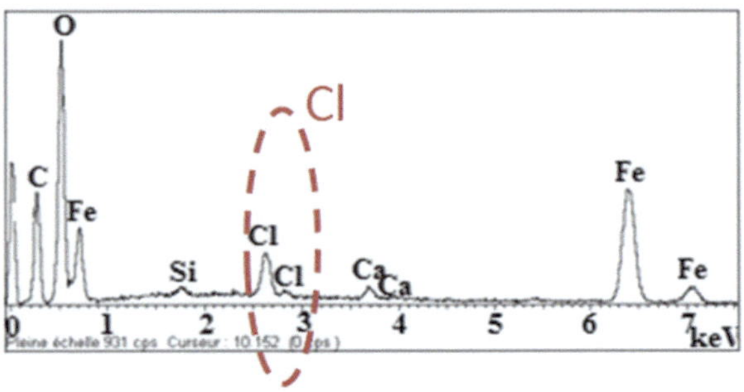

3.2 *Electrochemical Characterization*

Figure 8 presents the measurement results for the half-cell potential, the electrical resistance and the corrosion rate results at two different times. The first value (in light grey) is related to the corrosion state after curing (28 days curing + 3 days) while the second value (in dark grey) is related to the corrosion state after carbonation (24 weeks after curing including a month of preconditioning and 14 weeks of accelerated carbonation). From Fig. 8a it can be deduced that G1 specimens potentials measured after curing ranged between −230 and −579 mV with a mean value of −457 mV and a standard deviation (based on the 70 specimens) of 84 mV while after carbonation, the average value and the standard deviation were −550 and 14 mV, respectively (based on 57 specimens). This indicates that after carbonation, the potential values of the specimens rebar became more electronegative and showed less scatter. From Fig. 8b, after curing,, the average value and the standard deviation of the electrical resistance were 193 and 16 Ω, respectively, whereas after carbonation these values shifted to 97 and 13 Ω, respectively. The decrease by a 2 factor is rather unexpected as the carbonation process induces a reduction of the porosity. However both values (after fabrication and after carbonation) are of the same order of magni-

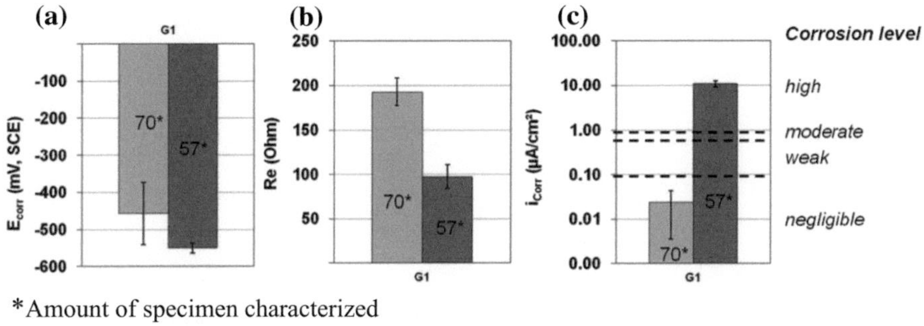

Fig. 8 **a** Half-cell potential, **b** resistance and **c** curent density of G1 specimens after curing (light grey) and after carbonation (dark grey)

tude (some hundreds) and usually a significant evolution for the reinforced concrete is rather a thousand. Finally, the corrosion rate after curing (Fig. 8c) was below 0.1 $\mu A/cm^2$ thus indicating a negligible level of corrosion according to Andrade et al. [9]. After carbonation a drastic increase of corrosion rate (values exceeding 10 $\mu A/cm^2$) demonstrated a high level of corrosion [9]. This could be explained by the release of the initially bound chloride ions induced by carbonation.

4 Electrochemical Chloride Extraction Treatment

ECE treatment deals with transformations of corrosion products, extraction of chloride ions, production of hydroxyl ions and migration of ions. This part of the paper aims to study these different issues.

4.1 Corrosion Products Characterization

Raman spectroscopy. During ECE treatment, regular Raman acquisitions (every day) were performed on a G1 specimen before carbonation to follow the phase transformations induced. After two days of polarization, the main phase detected was magnetite (main Raman band at 670 cm^{-1}) (Fig. 9). This phase remained stable during 2 weeks of polarization after which the treatment was ended. Further experimentations will be done for a Raman microspectrometry study on a chloride-contaminated specimen which is carbonated.

Scanning electron microscopy. SEM images after treatment (Fig. 10) showed no evolution of the corrosion layer thickness (Fig. 10a) and evidenced a morphological evolution of these corrosion products (Fig. 10b). Elemental analysis allowed verifying the absence of Cl in the corrosion products composition after treatment

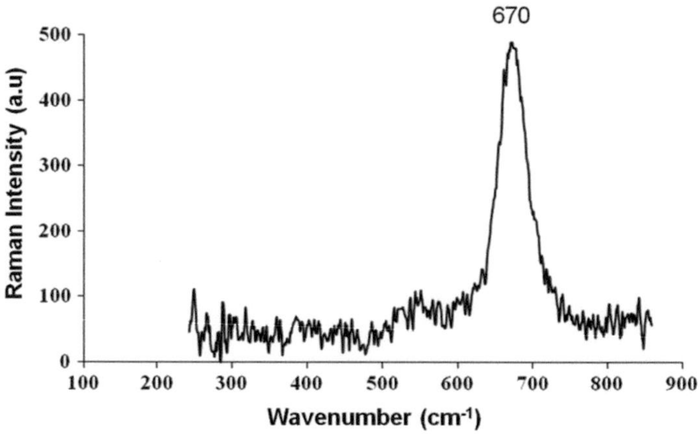

Fig. 9 Magnetite Raman spectrum

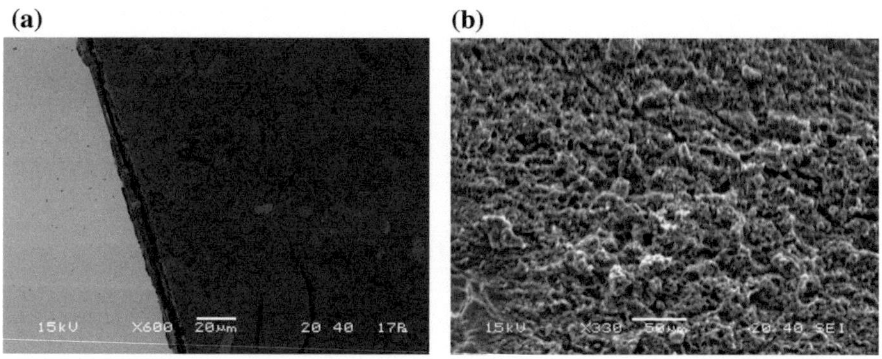

Fig. 10 **a** BEIC-SEM image (×600) of the steel/concrete interface on cross section of specimens type G1 after treatment, **b** SEI-SEM images of corrosion products observed into the rebar imprint on fracture after treatment ×330

Fig. 11 EDS elemental analysis of corrosion products after treatment

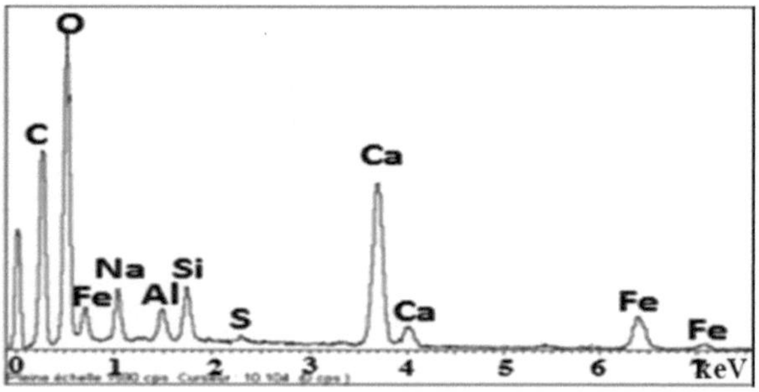

(Fig. 11). Other elements like Na and Al were detected into the corrosion products after treatment.

4.2 Chloride Titration

Figure 12 shows the evolution of chloride content in the concrete during the ECE treatment. The chloride threshold level for preventing corrosion in new structures as indicated in the EN206 standard [10] is also represented. Before treatment, the total and free chloride content was respectively 1.52 and 1.35% for the internal depth and was 0.94 and 0.94% for the external depth. These contents different from the 3% added in the mix were explained by a lixiviation during the curing (28 days in water). After 3, 7, 14 and 28 days of ECE treatment the chloride ions extraction amounted to about 52, 67, 89 and 92%, respectively, in the external cover and 27, 70, 92 and 96%, respectively, in the centimetre around the rebar. The extraction was faster during the first seven days of treatment which is in accordance with literature results [4–7]. Results show that the chloride content became less than 0.4% by weight of cement after 3–14 days of ECE treatment. Figure 12 also highlights that the difference between total and free chloride ions contents decreased with the duration of treatment which could mean that initially bound chloride ions can be electrochemically extracted.

For the non-polarized specimens, the chloride content decreased due to lixiviation but was still above 0.4% by weight of cement around the rebar after 28 days of immersion.

Finally, the results of chloride ions extraction obtained from the concrete and from the electrolyte were in good agreement.

Figure 12 also shows the evolution of alkalinity in the vicinity of the rebar. The observed realkalisation was probably due to production of hydroxyl ions around the rebar and to the migration of cations towards the rebar. Based on the obtained results and the standard [11] concerning realkalisation, the minimum treatment duration required is either 28 days to obtain a 10 mm realkalised concrete zone around the rebar or a range of 7–14 days considering the experimental rebar diameter (5 mm).

4.3 Evaluation of the Medium-Term Performance of the ECE Treatment

The durability of the efficiency of the ECE treatment depending on its duration (7, 14, 28 and 56 days) was studied 5, 10 and 15 weeks after the end of treatment (Fig. 13), following the corrosion rates of 20 specimens. The first five weeks were necessary for the steel rebars to depolarize after the treatment. Results show that 7 days of treatment ($i_{Corr} > 1\,\mu A/cm^2$) was not sufficient to halt corrosion. In order to obtain a negligible level of the rebar's corrosion rate ($i_{Corr} < 0.1\,\mu A/cm^2$), a minimum of 14 days appeared to be necessary. This duration is in good agreement with the duration for which the chloride content became less than 0.4% by weight of cement. Figure 13 also presents the corrosion rates of specimens in immersion for 56 days without current. Comparing the corrosion rates of the samples treated for 56 days ($i_{Corr} < 0.1\,\mu A/cm^2$) and the

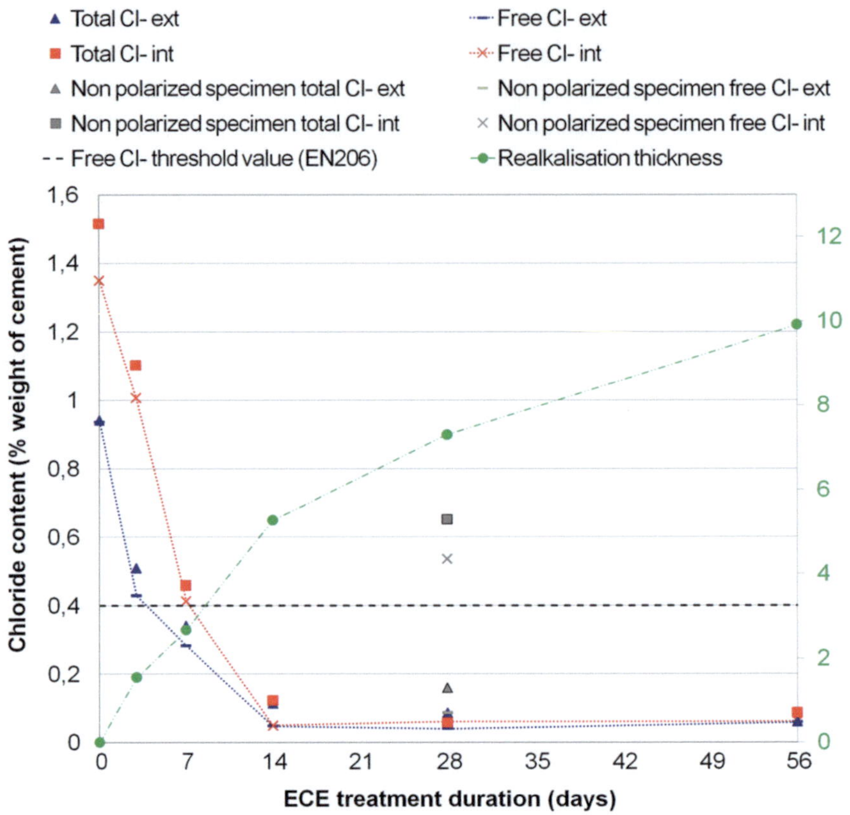

Fig. 12 Evolution of chloride content and of the realkalisation ring thickness around the rebar (pH > 9) during ECE treatment

samples only immersed for 56 days ($i_{Corr} > 1 \mu A/cm^2$), it appears that the latter are still corroding. This can be explained by the fact that although the chloride ions have been removed during the immersion, the realkalisation has not been performed and therefore corrosion remains active.

5 Conclusions

In this experimental work, the set-up designs adopted to study, first the corrosion processes in carbonated and chloride-contaminated reinforced concretes, and second, the electrochemical chloride extraction treatment were presented.

Results obtained on G1 specimens allowed to establish the following conclusions:

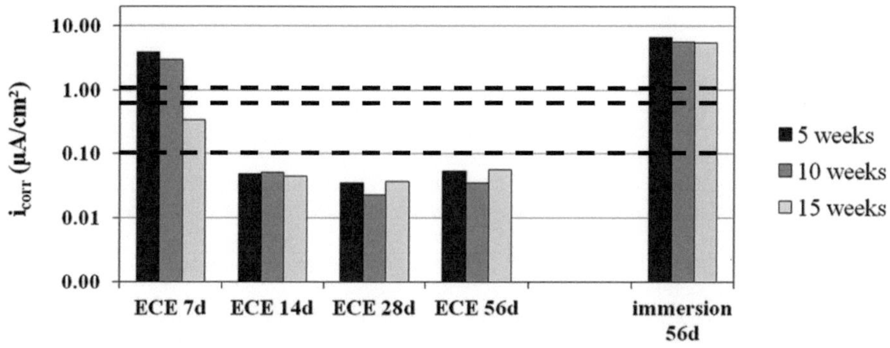

Fig. 13 Evolution of corrosion current density for G1 specimens treated (7, 14, 28 and 56 days) and immersed 56 days (non polarized), 5, 10 and 15 weeks after ECE treatment

- Chloride-contaminated G1 specimens showed a negligible corrosion level 30 days after curing as indicated by corrosion rates less than 0.1 µA/cm^2.
- After the carbonation aging, as expected, the corrosion rate increased rather drastically as evidenced by corrosion rates higher than 10 µA/cm^2. SEM results confirmed this behaviour with the observation of corrosion products on most of the interface (5–20 µm on the rebar and up to 250 µm of penetration within the surrounding the concrete).
- Efficiency of the electrochemical treatment based on chloride extraction as well as realkalisation of the concrete around the reinforcement was demonstrated. Two weeks of treatment were needed to obtain a free Cl^- content below 0.4% by weight of cement while 28 days of treatment resulted into one centimeter of realkalisation around the rebar.
- Electrochemical characterizations evidenced a negligible corrosion level ($i_{corr} < 0.1$ µA.cm^2) for a ECE treatment duration longer than 14 days.
- Finally, regarding the durability of the treatment efficiency, after almost 4 months, the corrosion rate remained at a negligible level (no resumption of corrosion).

The durability issue will be further studied by exposure to cycles of relative humidity at 20 °C (1 week 92%RH/2 weeks 60%RH) by means of electrochemical characterizations. Comparison with the 3 other series will give a larger view on this subject.

References

1. Tissier Y, Bouteiller V, Marie-Victoire E, Joiret S, Chaussadent T (2016) Corrosion processes and ECE treatment in a both carbonated and chlorinated reinforced concrete. In: Materials systems and structures in civil engineering conference, 21–24, August 2016. Lingby, Denmark
2. Tong YY (2009) Traitement électrochimique de réalcalinisation pour la réparation du béton armé dégradé par carbonatation, PhD thesis, Université Pierre et Marie Curie France

3. CEN/TS 14038-2 (2011), Electrochemical realkalisation and chloride extraction treatments for reinforced concrete - Part2: chloride extraction
4. Herrera JCO, Escadeillas G, Arliguie G (2006) Electro-chemical chloride extraction: Influence of C3A of the cement on treatment efficiency. Cem Concr Res 36:1939–1946
5. Swamy RN, McHugh S (2006) Effectiveness and structural implications of electrochemical chloride extraction from reinforced concrete beams. Cem Concr Compos 28:722–733
6. Fajardo G, Escadeillas G, Arliguie G (2006) Electrochemical chloride extraction (ECE) from steel-reinforced concrete specimens contaminated by "artificial" sea-water. Corros Sci 48:110–125
7. Arya C, Sa'id-Shawqi Q, Vassie PRW (1996) Factors influencing electrochemical removal of chloride from concrete. Cem Concr Res 26:851–860
8. Boucherit N, Hugot-Le Goff A, Joiret S (1991) Raman studies of corrosion films grown on Fe and Fe-6Mo in pitting conditions. Corros Sci 32:497–507
9. Andrade C, Alonso C (2004) Rilem TC 154-EMC Test methods for on-site corrosion rate measurement of steel reinforcement in concrete by means of the polarization resistance method. Mater Struct 37:623–643
10. EN 206 (2014) Concrete—Specification, performance, production and conformity
11. EN 14038-1 (2016) Electrochemical realkalisation and chloride extraction treatments for reinforced concrete Part1: Realkalisation

MIX
Papier aus verantwortungsvollen Quellen
Paper from responsible sources
FSC® C105338

If you have any concerns about our products,
you can contact us on
ProductSafety@springernature.com

In case Publisher is established outside the EU,
the EU authorized representative is:
Springer Nature Customer Service Center GmbH
Europaplatz 3, 69115 Heidelberg, Germany

Printed by Libri Plureos GmbH
in Hamburg, Germany